L K
FOR TH

LIVESTOCK
FOR A SMALL EARTH

THE ROLE OF ANIMALS IN A JUST AND SUSTAINABLE WORLD

Edited by Jerry Aaker
Heifer Project International

Contributors
James DeVries ▪ Dan Gudahl ▪ Jim Hoey
Robert K. Pelant ▪ Jennifer Shumaker

Illustrations by Barbara W. Carter

SEVEN LOCKS PRESS   WASHINGTON, D.C.

First printed June 1994

Library of Congress Cataloging-in-Publication Data

Livestock for a small earth : the role of animals in a just and
sustainable world / edited by Jerry Aaker ; contributors,
James DeVries . . . [et. al.] ; illustrations by Barbara W. Carter.
    p. cm.
   Includes bibliographical references.
   ISBN 0-929765-28-1
    1. Livestock. 2. Livestock projects. 3. Sustainable agriculture.
4. Farms, Small. 5. Heifer Project. I. Aaker, Jerry II. DeVries, James.
III. Heifer Project.
SF41.L58   1994
338.1′76′0091724—dc20   94-10382
                                             CIP

Manufactured in the United States of America
Cover and book design by New Age Graphics, Bethesda, Maryland

Seven Locks Press
Washington, D.C.
1-800-354-5348

This book is dedicated to those who have given so much of themselves to the vision of building a better world through the program of Heifer Project International. The founder of HPI, Dan West, provided the inspiration and the first practical demonstrations of how animals could make a unique contribution to the alleviation of hunger and poverty around the world. He and others set the course. It is left to our generation to build on the experience and learnings of the past 50 years in order to face the great challenges of our day, and to preserve the dignity of people and the integrity of all creation.

The lives and farms of hundreds of thousands of families around the world have been blessed by keeping and caring for farm animals. Perhaps more importantly, they have been further enriched by sharing their knowledge and offspring animals with neighbors through "passing on the gift." It is in that same spirit of sharing that we offer this book as a contribution to others who want to learn more about the role of animals in a just and sustainable world.

Jo Luck Cargile
Executive Director
Heifer Project International

TABLE OF
Contents

ix   Foreword

xi   Acknowledgments

xiii   Introduction

1   **PART ONE:
WHAT IS SUSTAINABLE DEVELOPMENT?**

3   CHAPTER ONE: The Benefits of Sustainable Animal Agriculture

13   CHAPTER TWO: Caring for Our Earth

27   CHAPTER THREE: Sustainable Agriculture: Humane and Socially Just

43   **PART TWO:
THE PROCESS MATTERS**

45   CHAPTER FOUR: People Will Say "We Did It Ourselves"

55   CHAPTER FIVE: Training for Transformation

67   CHAPTER SIX: Helping People Through Livestock Projects

79   CHAPTER SEVEN: Passing on the Gift

87   CHAPTER EIGHT: Behind These Animals Is a World of Hope

93   APPENDIX 1:   About Heifer Project International

95   APPENDIX 2:   HPI Cornerstones for Sustainable Development

99   APPENDIX 3:   The Accountability Process

103   Reference List and Bibliography

Foreword

'This book is a collection of essays that share what has been learned over the course of half a century about how to make agricultural development socially just and environmentally sustainable. It is a primer on development assistance using animals. The book is like a gift, in that it offers insights about programs that work. Perhaps an alternative subtitle could be "Sharing the gift of living and the gift of life." This would capture the essence of this work.

The book is useful for scholars, policy makers, development program officers, local community development workers, and farmers and their families. We all could benefit from better insights about the need for farm animals and animal welfare ethics. We need greater understanding about the complex dynamics of community development and self-help, particularly in situations of social marginalization. We need knowledge about the connection between sustainability and justice. It takes no stretch of the imagination to recommend to everyone that examining the long track record and learnings of an outstanding international rural development organization will be of lasting value. The stories from which the lessons are drawn are simple. The integration of what is written between the lines is where the wisdom is gained.

C. Dean Freudenberger
St. Paul, Minnesota

Acknowledgments

Fifty years of experience with the role of livestock in the developing world cannot be adequately encompassed in a single volume. Nor can we adequately recognize all the people who have shared their knowledge and wisdom with us over the years. Nevertheless, this book attempts to share some of the insights gained from Heifer Project International's experience in responding to human need through the use of animal agriculture. Many of the ideas expressed in this book are not attributed to particular individuals. Rather, the contributors, all of whom work in the International Program Department of HPI, wish to acknowledge that we have each benefited enormously from the insights of hundreds of colleagues and friends around the world. We especially want to recognize our indebtedness to the farmers we have the great privilege to work with in so many countries.

Special thanks goes to several people who reviewed and commented on the manuscript, including Darrell Huddleson, Rosale Sinn, Steve Muntz, Henryk Jasiorowski, Beth Miller, and Erwin Kinsey. Many of their suggestions are incorporated here. Also, thanks to Anna Zawada for her help in compiling the bibliography.

Each of our coworkers and partners in the dozens of countries and states where we have worked and traveled deserve to be named and recognized. However, at the risk of leaving anyone off this list, we simply want to express our heartfelt appreciation and encouragement to these colleagues. We commit ourselves to continue to dialogue and learn how we can work together for a more just and sustainable future for the world's children.

Introduction

While he was president of Tanzania, one of the poorest countries in Africa, Julius Nyerere wrote that development does not consist of more roads, more factories, more cars, or even more maize or food. In its most profound sense, development has to do with people—people changing to take control of their own lives. It is people becoming more human, more of what God intended them to be, willing to make decisions and take responsibility for their own destiny. It means people having access to the proper resources to implement decisions they make.

More recently the field of development has moved to a deeper understanding of ecology as one of its organizing principles. Without due consideration for the land and the integrity of creation, all the world's people are doomed. Ecology, not economics, must be the dominant principle in this new understanding of development.

We can apply the same principles to livestock development programs. More animals, animal products, or income from integrated animal agriculture is not the same as development. Animals may well be a stimulus to generate development, but in itself livestock production cannot be the primary goal of development or the only measure of its success. Development is often defined in material terms; but as several writers in this field suggest, this is not an adequate definition. It neglects the human component. Human development is a participatory process that leads to self-determination, self-confidence, mutual cooperation, and a better quality of life. Its goal is holistic transformation (DeVries 1992). A primary aim of transformation is to empower people and, in the process, to uphold the integrity of creation.

In this book we focus on three important elements to keep in mind when thinking about development based on sustainable agriculture: the land, farm animals, and communities. Obviously, successful farming requires other elements, including adequate capital to start farming and to continue to produce. The book discusses the various social, material, and natural resources required, and the threads of land, livestock, and people are woven throughout its pages. Our goal is to present a holistic approach to sustainable development. Our special interest is to integrate livestock into this tapestry. We would like to communicate this view to people who are concerned about poverty alleviation and sustainable development. These issues are becoming increasingly important as a growing world population puts ever-greater demands on the environment.

We believe a healthy environment is a prerequisite for people to have hope for the future. People are not isolated from an ecosystem. It is not possible to help people help themselves through livestock programs without considering their relationship to the various ecosystems in which they live. Such an approach would only be another example of what humans have done too much of already—tinkering with the parts instead of working with the whole. Humans are but one part of this whole, as are animals.

In the last decade a much better understanding has emerged of the important and unique role of livestock as a medium for human development. This book focuses on small-scale farmers in low-income countries, as well as in North America, and sustainability is seen from their perspective. Some who have power or economic interests in local communities, in whole nations, or in international structures may have conflicting viewpoints. Given the influence of agribusiness and the power of government to forge macrolevel policies for agriculture, we feel that small-scale, family-based farmers need more support in order to protect their interests, their land, and their way of life. The sustainable development of any nation cannot be achieved without a balance of interests that includes attention to this voice. Widespread poverty and sustainability cannot coexist.

The book is divided into two major parts. Part One attempts to answer the question, "What is sustainable development?" This question is addressed from the overall perspective of the importance of caring for creation and with a recognition that the natural resource base of the earth is under tremendous pressure. In analyzing and describing sustainable agriculture and the accompanying role and benefits of livestock, we use these criteria: Sustainable agriculture must be (1) ecologically sound, (2) economically viable, and (3) socially just and humane.

Part Two deals with the process of development, with particular attention to the participation of people in rural communities. We present several examples and stories of livestock development in grassroots projects in Africa, Asia, Latin America, and in rural areas of the United States. We look at how farmers with limited resources integrate livestock into their production enterprises and how an effective livestock development project works. What are some of the issues that marginalized rural people face in planning the use of their limited resources? What are some ways for small-scale farmers to set realistic goals? What about the values that underlie sustainability? How can we "enhance the dignity of all" through developmental assistance? We look at factors that people should consider when undertaking a livestock enterprise, including community and cultural factors that are such important components in this blend of people, land, and livestock.

Part Two also describes "passing on the gift" as a tool for sustainable development—the ingenious principle whereby those who receive help through livestock become donors to others. We conclude the main text of the book with some comments on visioning the future.

The appendixes present specific information about Heifer Project International (HPI). We include a description of HPI's "cornerstones," the fundamental principles upon which HPI bases its work around the world. Finally, an extensive list of reading references is provided for anyone interested in reading more about both the theory and practice of sustainable development. Many practical manuals, training tools, and reference books are listed for those who might want to study particular aspects of development, sustainable agriculture, planning, or the raising and care of various species of livestock.

B. Carter

PART

One

What Is Sustainable Development?

A sustainable future depends upon our commitment to creating an agricultural system that nurtures and beautifies the land, while providing health, abundant food, and a good living for farm families.

—LARRY OLSON, FARMER AND PASTOR

The basic premise of this book is that sustainable agriculture builds up and supports people, land, and animals. E. F. Schumacher observed that if you "study how a society uses its land, you can come to a pretty good conclusion as to what its future will be" (1973, 102).

Soil sustains life. Agriculture that does not care for the soil cannot hope to take care of people. In healthy soil, plants capture energy from the sun—energy that supports all life on earth, including people. Healthy soil that supports growing plants does not wash away with heavy rains; nor does it leach out or burn in the hot sun.

But is it enough to just take care of the soil? The *culture* in *agriculture* points us to the importance of sustaining human communities. We might say that agriculture that does not take care of the people cannot hope to take care of the soil.

People who are hungry may feel they cannot afford to leave land fallow or plant soil-renewing grasses, cover crops, or trees. People might not recognize that cutting down the rainforest or overloading the soil with chemicals will ultimately harm them, their children, and their grandchildren. One of the hopeful signs of our day is a raised consciousness about the importance of caring for the earth and the importance of sustainable approaches to agriculture.

More than a century ago Chief Seattle is reputed to have talked about the connection between people and the earth in this way:

All things on earth are connected. Whatever befalls the earth befalls the sons of earth. Man did not weave the web of life. He is merely a strand in it. Whatever he does to the web, he does to himself.

Most of this book deals with livestock and agriculture; but in the broader sense, it focuses on sustainable development. We believe that people have the capacity to make development sustainable. As stated by the World Commission on Environment and Development in *Our Common Future* (1987), sustainable development is "development that meets the needs of the present without compromising the ability of future generations to meet their own needs."

Sustainability is an ideal to strive for rather than a recipe to follow. Terry Gips (1986), of the International Alliance for Sustainable Agriculture, said that agriculture is sustainable if it is *economically viable, ecologically sound, socially just, and humane.* In this section of the book we look at these four elements of sustainable agriculture, with particular reference to how livestock fit into this definition.

Heifer Project International has found, after 50 years of working with livestock in development, that livestock can make positive contributions to the development of people as well as to the enhancement of the environment. In fact, in some situations sustainable agriculture may only be possible through a combination of animal and plant agriculture.

One

The Benefits of Sustainable Animal Agriculture

While visiting a family participating in a livestock project in Cameroon, a development worker saw a small boy, not over 10 years old, arrive with a huge basket of greens on his head. This family was earning a "full-time salary" from their livestock project. Over 95 percent of the rabbit feed on that small farm came from grass and farm byproducts that would otherwise be considered weeds.

This example demonstrates one strength of most projects in Africa—they involve the whole family. Everyone in the family not only helps with the work; everyone also benefits. Manure from the animals builds soil fertility and helps the family grow vegetables and grain for home consumption. The family consumes milk and meat and sells the surplus within the village or shares it with needy neighbors. One village leader remarked that before the project many children suffered *kwashiorkor* (severe malnutrition), but this has "now disappeared." Trips to the hospital are also much less frequent. Other obvious benefits include improved houses and money to pay school fees.

Philip Nju, the secretary of the Bamenda Dairy Cooperative in Cameroon, remarks that the most important aspect of their project is that people are working together. His dairy group last year organized a project to supply clean water to over 500 families. Another group recently opened a health clinic. Others have helped start a school, dug latrines, cleaned up the village, and started a credit association.

The projects in Cameroon have also generated a spirit of sharing. Recently one farmer lost all his cows to disease. Another farmer was ready to pass on the gift to him. "My cow was pregnant, and I told him if she produces a heifer, it's yours. Lucky enough for him, it was." The project team did not need to introduce the spirit of sharing into this culture, but they have certainly encouraged it.

The above story illustrates how animals provide people with benefits that are not purely economic. Another example is planting grasses and trees, a practice that fits well into a livestock production system and also provides erosion control to help conserve natural resources. Perhaps the most unique role of livestock in a traditional farming system is to minimize risk. Animals have two flexible assets that increase stability. First, animals can move to find food, water, and shelter, as well as to avoid danger. Second, they require no rigid timetable for harvest or breeding. In good times animal numbers can be expanded to store extra food or capital (the living bank account). In bad times these animals can be sold or eaten. During periods of economic instability with high inflation, livestock can take the place of both the savings and the checking accounts. A paper published by the Worldwatch Institute (1991, 8) gives the example of a six-year study in Lesotho, which found that farmers investing in cattle earned 10 percent interest, while bank accounts over the same period lost 10 percent to inflation.

Livestock production for human consumption usually leads most people to think immediately of meat. However, livestock may enhance the diet in a variety of ways other than meat consumption. They can provide draft power to till the land for grain production, manure to fertilize a garden plot, offspring to sell in order to buy cheaper staple foods, gifts for special occasions, milk to share with malnourished children or to sell, cheese for home consumption or marketing, eggs for protein or for sale, honey for sale or sweetening. Pigs can serve as a savings account during periods of food shortage and also as garbage disposers.

The benefits that livestock give to people can be thought of as the seven *M*'s:

- Milk
- Money
- Meat

- Manure
- Movement
- Material
- Motivation

Milk—Nature's Most Perfect Food?

Milk is the food that all mammals depend on for survival at birth. It is nutritious and a good source of calcium, riboflavin, and protein. About 75 percent of the calcium in the United States food supply comes from dairy products. Some cultures, such as the Maasai or Fulani of Africa, depend on cow's or goat's milk for sustenance throughout their lives. It is a part of their culture.

Breast milk is definitely best for infants. Health and development agencies concerned with childhood health and nutrition recommend that infants be breast fed during the first year of life or longer. But cow's or goat's milk is an important food for growing children. While some pediatricians have argued that nondairy food sources can provide adequate calcium, a look at the list of the alternatives makes some parents skeptical. For example, 300 milligrams of calcium can be supplied by one cup of milk; one-and-a-half cups of dark green, leafy vegetables (spinach, for example); eight ounces of tofu; or two cups of cooked soybeans. The daily dietary need for children and adults is about 800 milligrams of calcium. Pregnant and nursing women and adolescents need 1,200 milligrams. Clearly, parents should try to provide a balanced diet, in which children obtain calcium from both milk and vegetables.

Some people develop a sensitivity to milk called *lactose intolerance*, which causes digestive discomfort. A person with lactose intolerance has difficulty digesting one component in milk—its sugar. This sugar, called *lactose*, is normally broken down by an enzyme made in the body called *lactase*. This sensitivity can often be avoided through consumption of fermented milk products such as cheese, yogurt, or soured milk. In these products, a bacterial process changes the allergen, lactose, to a more digestible product. In addition, cheeses usually do not cause symptoms because most of the lactose is removed during processing. Lowfat or skim milk is also suggested, but in small servings. Also, goat's milk can be a recommended substitute for cow's milk for children or adults with digestion problems. Lactose intolerance usually starts in late childhood, and it affects some ethnic groups more than others. Most children can drink milk without problems.

Raising livestock for milk production is a cost-efficient means to provide essential nutrients. In the Western world, where beef cattle are increasingly produced in an industrialized, grain-based system, the cost is high in terms of grain per pound of meat. One pound of beef requires about 16 pounds (7 kilograms) of grain. For this reason, grain-fed beef is essentially a dietary component of the world's wealthiest societies. But R. E. McDowell (1991) notes that in less industrialized countries,

Milk goats at the Karen Training Center in Thailand have introduced many people to goat's milk. People who never thought they would like any kind of milk, especially goat's milk, find it palatable, and they feel better after drinking it. In villages where meat goats were introduced, the diets are slowly changing. People who first thought they would not like goat meat soon found they did. This has reduced their dependency on wild game. The use of manure was a bonus. It is mostly used on gardens, fruit and coffee trees, and pasture grass. In the village of Huey Hawn it is also used to fertilize a fish pond near the goat pen.

—Report from
Karen Goat Project,
Thailand

1 kilogram of grain fed to a dairy cow, when coupled with forages, provides 6 kilograms of milk. In low-income countries, small-scale farmers commonly keep crossbred cows, which can produce up to 16 liters of milk a day from grasses and leguminous plants. Also, the term *grain* usually refers to concentrates, which are made up of a high percentage of agricultural products that would be wasted unless fed to livestock.

The demand for milk in developing countries exceeds the supply. Nonfat dry milk, or "filled" milk, is imported from North America, Europe, and Australia. Even in countries where many adults are lactose intolerant, milk is in demand as the most effective means of alleviating malnutrition. A study done by a graduate student at the Zamorano Agricultural College in Honduras demonstrated that the level of malnutrition in a group of Honduran children dropped from 90 percent before the introduction of goat's milk to 20 percent after one year of dairy goat development in the area. Health clinics in areas with high incidences of malnutrition routinely provide powdered milk as a protein supplement to lower the level of malnutrition in children.

If a diet includes a small amount of good-quality protein, large amounts of carbohydrates can make up the rest of the diet. According to McDowell (1991), one-third of a cup of milk (or 25 grams—one-twentieth of a pound—of meat) combined with two-and-a-half to three pounds of cereal or tubers per day meets the human requirement for protein. Milk also improves the digestibility of cereal grains. Without milk, less than 30 percent of the cereal protein is used for growth; with milk, over 60 percent is used.

Money—"Almost" Everybody Needs It

Many rural people do not have access to banks or credit. Because of currency instabilities, inflation, or inaccessibility, the bank may not be the first choice for investment of money. In many cases, farmers prefer to keep their wealth in the form of animals. Animals grow and reproduce (a kind of interest on investment). Many farm animals provide regular income through milk, egg, or honey sales. In short, animals can act as a mobile and living bank account.

Although most of the world's people operate in a money-based economy, a significant number of the world's poorest people still function in a bartering system. Animals and crops may be traded with neighbors and at local markets for other basic needs, such as kerosene or salt. In traditional agrarian societies people commonly bring pigs, chickens, and goats to the weekly market to sell for money or to trade directly for other needed goods.

Meat—Is It OK to Eat It?

Many kinds of animals are raised or hunted for meat. Meat is an important part of the diet of people in most cultures in the world. People generally enjoy the flavor of meat—from iguana to mopani worms, beef to guinea pigs and poultry—even though cultural preferences vary. Meat is a concentrated source of protein, minerals, and B-complex vitamins.

Most meals for the affluent of the world are meat-based; but for the poor of the world, meat is a luxury food. It is frequently used as a condiment rather than as the central portion of a meal. Meat is often too expensive for poor families. Smaller animals, such as guinea pigs or rabbits, can provide an ideal alternative for resource-poor farmers, especially if refrigeration is not available. Perhaps the most universally popular meat source is poultry. The advantage of both smaller animals and poultry is that they can be consumed at one meal.

Another advantage of meat is that it can be kept for long periods of time in the form of a live animal without the risk of spoilage. Often when beans or grains are stored over long periods, the postharvest losses from insects, vermin, or improper storage can be considerable. As long as an animal is healthy and well cared for, storage "on the hoof" is not a problem, even though it can be a cost.

The use of meat for human consumption has come under attack from many directions. Environmentalists, animal rights activists, and health advocates all point to meat consumption as a culprit for a variety of today's problems: deterioration of the global ecosystem, ethical problems exemplified by abuse of farm animals, and degenerative diseases like atherosclerosis. Even global poverty and hunger have been blamed on the meat-based diet of the rich. Thus, some critics believe, elimination of meat from the human diet would solve most of the world's food problems. Some seem to want to stop all meat production and raising of animals.

Studies have shown that the quantity of calories consumed levels out, and people stop increasing their food intake at a certain income level. But the type of food eaten changes dramatically as income rises. Cassava and other low-quality root crops are replaced by other staples with higher nutritive value, such as beans, rice, and corn. A low-income rural family will usually sell its livestock products to buy cheaper grains and beans to eat. People with higher incomes consume more expensive animal products, such as milk, meat, and cheese.

People in the United States and other wealthy nations often eat far more animal products than are necessary for adequate protein. In these populations, people eat meat purely for the taste. On the other end of the scale, many families in developing areas have not yet reached the level of adequate protein intake. An optimal level of meat consumption lies somewhere between these two extremes.

The impact of meat consumption on human health differs between wealthy and poor countries. High meat consumption in the United

States has been blamed for degenerative diseases such as heart failure and cancer. But the addition of small amounts of animal products has a dramatic positive effect on the health of poorly nourished children. Certain alternative sources of protein, including good-quality legumes such as peas and beans, grow poorly in acidic tropical soils. Furthermore, these products cannot easily be stored over the dry season without severe losses. Low-quality starch foods that are the staples of most rural poor in the tropics—for example, cooking bananas, cassava, and other root crops such as yams and cocoyams—are deficient in an essential amino acid called *methionine*. The meat from one small chicken will supply both the daily protein and methionine needs for a typical family.

The type of animal product that is appropriate depends on the situation. Diseases associated with animal fats in the United States have multiple causes, though fat is, no doubt, a culprit. People in the United States have plenty of energy sources and need to limit their consumption of animal fat. In poor countries where fat—essential in small quantities—is often totally lacking in the diet, whole milk, eggs, and fatty meat may be nutritionally sound.

While limiting animal fat is certainly advisable, most diseases cannot be blamed exclusively on the consumption of meat. Marvin Harris, an anthropologist, points out in his book, *Our Kind* (1989), that Eskimos who live entirely on fatty animal products do not show high incidence of heart diseases. And archeological evidence indicates European races

in cold climates were almost exclusively meat eaters before the beginnings of agriculture, and their height and bone quality decreased when they settled and started eating mostly grains. Height and bone quality improved when animals were domesticated and diets became varied once again.

Manure—Multiple Benefits

Besides providing the economic benefits of food, fiber, and cash, animals also produce manure, which can be valuable in supplying fertilizer or generating biogas for fuel. All animals produce manure and urine. Manure can be used directly on fields for fertilizing crops, or it can provide an important ingredient for making compost. Compost is especially effective in providing nutrients to plants and in building up soil. Manure can also be added to fish ponds to make algae grow, providing food for fish such as tilapia. Manure can be used as food for worms and is important for manufacturing methane gas in a biogas digester. In some countries, manure may be the only reliable source of fuel for cooking. It is collected, dried, and sold for this purpose. Some cultures even use it as a component in plaster for house walls and for medicinal purposes.

Movement (or Muscle)—Animal Power

When animals move and transport implements, people, and materials, they provide draft power. This can translate directly into economic benefit.

Tomasa Ellao is a member of a women's group in the KASAMA carabao (water buffalo) project in the Philippines. She commented on the importance of this draft animal to her and the family as a means to improve their lives by farming:

> Right now it is a big help to our family, particularly in the plowing of the fields, which is the source of our living. Instead of us renting a carabao at P100 a day (about U.S.$4), we now use our own carabao for plowing. The money which we would use for renting an animal now serves as additional funds for our everyday expenses.

Animals ease human burdens in a number of ways. They haul loads and pull carts and agricultural implements. Animals help lift water for irrigation and thresh grain by trampling on it. They also are still used extensively in many parts of the world as pack animals and to carry or pull people in wagons from place to place, either out of necessity or for recreation. Horses, oxen, donkeys, water buffalo, and camels (as well as llamas in South America) are the most commonly used species for these purposes. Animals obtain energy from grass and other renewable plants and do not consume nonrenewable fuels such as gas or diesel fuels. Obviously, they do not need spare parts, as do cars or tractors. In fact, they provide their own replacements.

In 1991 a development organization in Zimbabwe initiated a donkey power project in a resettlement area. The people in this area could not afford to send their children to school. About 75 percent of the population was illiterate. Diet was generally deficient in protein and consisted mostly of maize (corn), tuber plants, or sorghum. Cotton, corn, and sunflowers were also grown for income. Some people in the area owned a few chickens, goats, or sheep; but the overriding problem was a lack of draft power to till available land. Malnutrition and lack of income were common. Surveys showed that 57 percent of the children were malnourished.

To break the cycle of poverty and disease, the development organi-

zation organized supplementary feeding programs for mothers and children, especially during the months when food was scarce. This was a good stop-gap measure, but it did not solve the long-term problem.

As a pilot project, donkeys were provided to farmers in this area in 1987. For example, Mr. Sibanda, husband of 3 wives and father of 13 children, had 12 acres of land. He received a set of harnesses, a plow, and 4 donkeys. Prior to this he could only produce a few bags of corn and sorghum, a few groundnuts, and one-and-a-half bags of cotton. With draft power, the Sibanda family produced 35 bags of corn, 12 bags of sorghum, 6 bags of groundnuts, and 9 bales of cotton. In addition, Mr. Sibanda's donkeys helped him transport produce and water, made plowing easier, provided manure, and increased his prestige in the community. Other farmers in the area who received draft donkeys had similar experiences.

After this successful pilot project many more families joined the draft donkey association. Having a larger harvest has meant more to eat, less malnutrition, and more income to pay for school fees.

Material—Renewable and Essential

Wool and mohair are renewable resources. Fiber, together with hides, skins, bone, and horns, are a few of the material byproducts animals provide. Humans have depended on these materials for thousands of years for shelter, clothing, tools, and art work. Native peoples in North America were astounded to see white hunters slaughtering the bison of the Great Plains for the mere pleasure of the hunt. For the Native Americans the bison were friends that provided shelter, clothes, food, and tools. In many cultures today, animals continue to give this same full measure of material benefit to people. In the Andean region of South America, people depend on llamas, alpacas, and sheep for clothing from head to toe.

Motivation—Fulfilling Dreams

Perhaps the most important benefit of animals, from a holistic viewpoint, is the motivation and dignity that the ownership of an animal provides. Owning an animal, such as a milk cow, would be beyond the wildest dreams of subsistence farmers in many low-income areas. When farmers are able to own and care for an animal or two, and furthermore, help others through passing on the gift of livestock and training, a spirit of sharing contributes tremendously to a sense of dignity.

Being successful in a livestock enterprise, no matter how small, provides immeasurable encouragement for farmers who live in areas with limited opportunities. With the hope that an animal provides through milk, money, and manure, families and other groups are encouraged to pursue new goals and improve living standards. This motivation is often expressed in terms of the hopes they have for their children.

Beyond the Seven M's

Of course, the roles animals play are not limited to these *M*'s. Bees produce honey and pollinate crops, increasing yields. Poultry produce eggs as well as meat. Animals play an important role in keeping the farm in balance. Earthworms ingest soil and small rocks and leave castings in their wake to enrich the soil. Chickens consume bugs and weeds around the farm and provide high-nitrogen fertilizer. The family dog may serve as a burglar alarm, and the llama or donkey may protect other livestock from predators. The cat helps to keep the rodent population in check, as does the guinea hen. Wild birds may help eat garden insects, and the rooster provides a free wakeup call each morning.

Animals eat scraps and plants that humans would not relish eating. When managed properly, hoofed animals can have a restorative effect on pastures. Pigs can be used to dig up fields. Animals are used to harvest crops or glean fields after harvest. They may also provide a means to enhance the value of crops such as corn, especially if transportation of heavy, bulky grains is difficult and expensive or if markets are scarce.

Clearly, livestock provide people with many economic and material benefits. There are other benefits as well, many of them intangible. A holistic view of people, land, and livestock leads one to focus on hope, dignity, and social transformation. In the final analysis, these may be the more important benefits.

Two

Caring for Our Earth

*We must do more than simply make
a resource last as long as possible.
An ecological ethic will seek not only
to preserve the land but to enhance it.*

—DEAN FREUDENBERGER

Animals have a role to play in sustainable agriculture. We agree with the authors of *Farming for the Future* that ecologically sound development "is best ensured when the soil is managed and the health of crops, animals and people is maintained through biological processes. Emphasis is on the use of renewable resources" (Reintjies et al. 1992, 2).

The Solar Connection

For agriculture to be ecologically sound, energy must not be wasted, and the source of energy should be nonpolluting. The origin of all our energy (fossil fuels, hydropower, wind, biogas, solar power) is the sun. The sun causes the wind to blow, water to cycle, and, most importantly, plants to grow. To make the best use of the free energy from the sun, a farm needs healthy soil, a variety of plants (crops, forages, and trees), and different kinds of animals to turn green plants into fertilizer. Healthy soil with healthy plants captures and holds the sun's energy efficiently. But it does not stop there. Healthy plants benefit from animals. Some plants are eaten by animals to recycle the energy and nutrients in the form of meat, milk, and manure. Animals also move seeds from place to place by consuming and excreting them. Living creatures pollinate plants. In certain climates, animals must trample the dry grasses to the ground to decompose them and feed the soil. Sharp hooves can even help in the germination of seeds.

Besides recycling energy and helping spread seeds, animals provide other types of energy to people on the farm. Draft animals move both people and farm produce to market. These same animals plow the fields and provide energy to pump water and grind grain. More and more people in developing areas are learning to make biogas from manure. This gives families both the gas for cooking or lighting, as well as the slurry for fertilizer.

Talking about the role of livestock in a whole system actually means talking about all sorts of animal life. These include microscopic or larger insects, birds, reptiles, small rodents such as mice, and a host of other mammals. Some animals eat only plants, digesting those parts that are inedible by other animals. Some are predators and hunt other animals, keeping the number of plant-eating animals in check. Without these predator animals, plants can become overgrazed and less of the sun's energy is captured. Plants and animals evolved together in a symbiotic relationship. With less energy captured by plants, the whole system starts to disintegrate, as happens in desertification.

Of course, people cannot be separated from an ecosystem. The ecosystem on each farm keeps a "whole" system going. As managers of the farm ecosystem, humans can take the place of the larger predator animals that keep the natural systems in balance. It is not possible to help people help themselves with livestock programs without considering their relationship to the various ecosystems in which they live. Failing to understand this results in tinkering with the parts instead of working with the whole.

The Farm as Ecosystem

Human beings invariably create a different ecosystem through agriculture, because all agriculture is really just an intervention in nature. But it is an intervention that can be regenerative if done properly. The beginning of agriculture 10,000 years ago can be pinpointed as the beginning of the process that may end in the extinction of topsoil. The history of agriculture, viewed as a whole, stands as a classic study of what happens when humans do not pay attention to the way nature organizes itself.

Those who would promote and support livestock development projects must understand the ecosystem in which the project takes place. In many instances the original ecosystem has been drastically altered by human mismanagement. In such cases, a sustainable approach to agriculture is much closer to nature than what it is replacing.

Building a just and sustainable world involves the study of ecosys-

B. Carter

tems just as much as it involves the study of human political and social structures. Whenever we talk about an ecosystem we automatically include human beings. So it is reasonable and wise to look upon a farm as an ecosystem and to encourage farmers to do so. The farm as ecosystem is part of the larger system. If there is a unifying concept in a healthy ecosystem it would be diversity, and a key component of diversity is livestock.

Biodiversity

Nature is neither static nor constant. Diversity is the most important principle in ecology. Diversity allows ecosystems to adapt to the inconsistencies of climate, predator attacks, and diseases. Not all species of animals and plants will thrive at a given time, but some will. Depending on circumstances, different plants dominate a particular space.

On a global scale, humans have been reducing this natural biodiversity for some time. Genetic diversity is a nonrenewable resource. In the humid tropics, the loss of species in rainforests has become one of the most urgent environmental problems on the human agenda. At risk is the loss of the ability to find cures for diseases and genetic stock for new foods. But more serious for the planet is the loss of the earth's ability to adapt to changes, not to mention the biological and ethical implications of destroying other species for financial gain.

The threat of mass species extinction has been widely reported in the press. Scientists estimate that destruction of the tropical rainforests alone could wipe out up to a million species by the year 2000—about 100 species a day.

The rate of extinction has been increasing rapidly. Between the years 1600 and 1900, humans killed off about 75 species—1 every 4 years. In the next 90 years, another 75 species became extinct—a little under 1 a year. Paul and Anne Ehrlich (1981) wrote this about species extinction:

> Only in about the last half-century has it become clear that humanity has been forcing species populations to extinction at a rate greatly exceeding that of natural attrition and far beyond the rate at which natural processes can replace them. . . . The rate has been estimated to be between 5 and 50 times higher than it was through most of the aeons of the evolutionary past . . . in the last decade of the 20th century that rate is projected to rise to some 40–400 times "normal."

On the local scale, agriculture puts the control of genetic resources in the hands of the individual farmers. Agribusiness managers have trusted modern technology to overrule the principles of ecology. Chemicals for pest or disease control and irrigation to overcome unpredictable weather have made high productivity the primary concern of farmers. While productivity is always a primary concern in farming, preservation of species and conservation of soil must also be priorities.

Practically speaking, the healthiest soil is biologically diverse. Three ways to provide organic matter to the soil are composting, cover crops, and crop rotations. Applying relatively small amounts of compost to soils inoculates the soil with microorganisms and supplies it with nutrients and digested organic matter. Compost piles teem with microbes that serve several purposes. First, they "mop up" free nitrogen in the soil, incorporate it into their bodies, and then release it as they die. Second, microbes produce organic compounds that glue soil particles together. And third, the microbes decompose organic residues from previous crops.

Why Preserve Diversity?

A paper by the Institute for Agricultural Biodiversity in Decorah, Iowa (1992) points out that in a biological system, alternatives vanish as complexity declines. As variety increases, options multiply. In a changing environment, life needs a spectrum of possibilities in order to endure. *Sustainability demands diversity.*

In the genetic domain, the abundant crop and livestock varieties available today must be preserved to provide genetic possibilities for an unseen future. The maintenance of pure genetic types within healthy populations and the availability of this genetic information to everyone are the central concerns.

Each variety or breed expresses particular responses to unique problems. Climate tolerance, disease and pest resistance, and other traits are all genetically transmitted. These adaptations are the raw materials for meeting the challenges of an evolving agricultural environment. To allow them to become extinct is to invite disaster.

Why Agricultural Diversity Is Vanishing

The disappearance of agricultural diversity can be traced to a number of causes:

- The number and type of farms decreases daily.
- Standardized markets demand greater uniformity from producers.
- Markets and technologies put pressure on wildlife habitats.
- New plant and animal patenting laws may increase profitability for private developers, but they also narrow gene pools and concentrate them with a few owners.
- Vertical integration of agricultural genetic stocks, production inputs, and markets tends to reduce diversity throughout the system.
- Production and profit demands often overlook the need to preserve traits that may have no value in the present-day market but that may hold the key to survival in an uncertain tomorrow.

All these factors result in the erosion of biological and genetic diversity. The future of food production in a changing world depends upon preservation of this irreplaceable treasure trove.

Going Against the Grain?

Genetic diversity on many modern farms is a low priority. Monocropping of high-yielding varieties has spread to agriculture throughout the world, especially among farmers who can afford the costly inputs. Chemical fertilizer eliminates the need for livestock to turn residue into manure.

But the rules of nature cannot be so easily ignored: insects and microorganisms can develop resistance faster than technology can develop new chemicals to control them. Savory (1988) has described how spraying grasshoppers with pesticides results in successive generations of survivors that are ever more pesticide-tolerant; this buildup of resistance may occur as well in hundreds of untargeted species, such as mosquitoes. Thus, the introduction of chemicals often leads to an increase rather than a decrease in pest and weed problems. Irrigation has lowered the water table in many places. Chemicals have polluted groundwater and destroyed benevolent insects, worms, and soil organisms. Loss of organic matter in the soil has led to serious erosion problems and lack of water retention. The separation of crops and livestock has encouraged continuous monocropping of erodible farmland and has created factory farms for livestock.

It is true that science and technology have enhanced agriculture in ways that no rational farmer will completely reject. Plant breeders, for example, have developed some disease- and insect-resistant varieties, such as nematode-resistant soybeans. We are not arguing here for the rejection of technology, but for a balanced view that puts a high value on the health of people and the conservation of natural resources, especially soil and water.

Some farmers have not totally accepted industrial-agricultural technology and are attempting to return to more sustainable practices. Many can't afford the increasingly expensive package of fertilizers, pesticides, and irrigation. Traditional farmers understand from generations of learning that a farm can thrive through unpredictable weather and predator activity if enough diversity is present. Diversity on the farm means different types of livestock or crops as well as different varieties of the same species.

The following description of a diverse small-scale farm comes from a "poor" area of Appalachia in the United States. It may sound to some like a kind of back-to-the-land romanticism. In reality, it may point in the direction of a more sustainable life-style than is lived by much of the population in our economically "rich" country.

Diversity on a Small Farm in Kentucky

For generations the farmers in eastern Kentucky have relied on tobacco for their income. Now, because of reduced demand for tobacco and the recognition of its harmful effects on humans, some of the people in this area are involved in cooperatives whose shared goal is to find alternatives to dependency on this crop. Diversifying from tobacco is a long process, and livestock can be a part of that process.

Neil and Denise Hoffman are leaders in the Owsley County Cooperative Livestock Improvement Program (CLIP). The Hoffman farm occupies about 60 acres of hillside, plus some hilltop pasture. In addition to finding alternative row crops to replace tobacco, the Hoffmans now keep hogs and strawberries as their main income generators, with shiitake mushrooms and blueberries showing promise as well. Raspberries, green peppers, winter wheat, forest products, milk goats, bees, and chickens add diversity and income to the system.

The key to the Hoffmans' success can be found in the way they have integrated the different operations and used natural biological processes to assist them in their management. Bees ensure good pollination of the fruit crops and provide honey as a bonus. The Hoffmans use a draft horse to cultivate between the rows of crops, and Nubian milk goats produce milk for the family, the neighbors, and the hogs. A team of horses helped to drag stones and logs for the log cabin they built themselves.

Hogs are rotated around the farm, feeding on the lush hilltop pasture of orchard grass and lespedeza when it has gone to seed, so that the pasture reseeds itself every year. These same hogs feed on rape after it has served as the spring cover crop in the tobacco/green pepper field. In fall this field is planted in rye or winter wheat. Megan, the Hoffmans' 12-year-old daughter, has learned to find the pupal stage of the praying mantis in the forest, and she brings these pupae home to help in pest management.

In summary, this family has the sort of farm that both animal welfare groups such as the Humane Society of the United States (HSUS) and environmentalists such as Wendell Berry would commend. The sows are so healthy that Neil Hoffman was able to turn down a donation of antibiotics. He has no need for them in his operation. Except for a short period immediately after farrowing, the sows roam freely between their clean hoghouse and the shady woodland alley leading to the hilltop pastures that provide their main source of nutrition. The rape cover crop provides extra nourishment for the sows before farrowing in the late spring.

Generating all their income from the small farm, the Hoffmans have indeed set an example in their community and for struggling family farmers all over the United States. It is hard work, but their reward is good food, a healthy lifestyle, and the satisfaction of helping to improve their surroundings.

Diversity in Guatemala

It is important to keep in mind that the above example from the United States would be considered luxurious in many third world settings. The well-being of animals must always be seen in the context of overall living conditions. While recommending constant supplies of clean water and weatherproof shelters for animals, we realize that many third world rural families do not enjoy such luxuries themselves. These families would feel privileged to live in zero-grazing shelters, with partly cemented flooring and a constant supply of food and water.

In this context, a farm such as that of Feliciano Orellana in Guatemala is impressive in terms of integrating family, crops, and livestock. Feliciano's two goats are housed right beside the house for protection and accompany the family to the corn and bean fields every day to feed on grasses, leaves, and wild plants. Free-ranging chickens clean up after the family has eaten and keep the farm free of flies and insect pests. Two cows provide milk as well as manure for compost. This compost is being used to start a community tree nursery that provides both fruit and nitrogen-fixing forage trees.

The best fertilizer on the land
is the footsteps of the owner.

—ARISTOTLE

Indigenous Knowledge and Genetic Resources

Development agencies in industrialized countries must relearn the wisdom of diversity. This could mean changing their technical training to better match the biological reality of small-scale farming. For example, a World Bank technical paper recommends undertaking long-term research in Asia that goes beyond rice-based farming and integrates livestock and crop production (Barghouti et al. 1990). We know of many small-scale farmers in Asia who are striving for this kind of integration: farmers in Thailand, the Philippines, Indonesia, and China are working with water buffalo, fish, pigs, goats, cattle, and other animal species. Through sustainable livestock projects farmers are adding diversity to farming at the local level.

The so-called green revolution introduced high-yielding crop varieties that could be duplicated all over the world, decreasing genetic diversity on the farm. The primary focus of the green revolution was on crops and plants. Without the wealth of genetic material found on traditional farms from which new cultivars can be bred, whole species of food crops could be lost to disease. All the arguments applied to crops can be used to justify preserving traditional breeds of livestock as well.

Traditional management of genetics in livestock focuses, above all, on survival of the animal. Where livestock ownership is a source of pride and wealth, as in many African cultures, survival is a far greater force for genetic selection than production of food and fiber. Animals that are diseased, weak, or have poor mothering qualities will be sold or eaten, while those that have proven resistance to disease and drought are retained. In pastoral systems, animals that cannot walk long distances or that resist herding will be culled.

Unfortunately, subsistence farmers' need for cash sometimes forces them to sell off some of their stock that brings the best price. They may sell larger and more productive animals while keeping some genetically inferior stock for breeding. This occurs, for example, among the llama raisers of the Andean region of South America. In many places in Bolivia and Peru, where llamas and alpacas originate, genetic degradation has been observed as a result. The same thing has happened in the Philippines with water buffalo.

In the United States another driving force has been in effect: maximization of production. Some traditional breeds are no longer being raised because of the trend toward specialization.

When the goal of development is to improve nutrition and living conditions for farm families, animal productivity takes on great importance. In these situations, finding good-quality, locally grown feeds can produce the desired results. Better nutrition will improve productivity of local breeds and eliminate the need to import exotic breeds. Alpacas, guinea pigs, llamas, yak, alligators, some indigenous breeds of water buffalo, goats, rabbits, pigs, and cattle all respond to improved nutrition. Even if better nutrition improves productivity, however, going outside of the animals' localities for new blood lines needs to be considered to avoid inbreeding. Also, crossing unrelated animals can bring about the well-known "hybrid vigor" in offspring.

Local Breeds or Exotics?

Besides preserving genetic diversity, the use of traditional breeds is preferable for other reasons. Local livestock are often resistant to or tolerant of insect-borne tropical diseases that are so devastating to certain livestock, such as trypanosomiasis and East-coast fever in African cattle. Modern veterinary drugs can prevent some of these diseases in exotic, imported animals, but the use of imported or costly vaccines to prevent diseases cannot be viewed as sustainable.

Because animals build up resistance to certain medicines, such as *acaricides* (de-tickers), the lifespan of these medicines becomes increasingly shorter as they are used more intensively. *Ethnoveterinary medicines* (medicines and remedies produced from indigenous ingredients) are not only less expensive and renewable, but are also more consistently accessible in countries where politics and economics are unpredictable. The traditional knowledge that goes into the production of such medicines is in as much danger of extinction as the traditional breeds and should be preserved whenever possible.

Ideally, local breeds should be used in livestock development programs. Frequently farmers in low-income areas ask for help to get exotic breeds for both production and status reasons. Crossbreeds should be recommended to these farmers. Not only are such breeds more resistant to disease than pure exotic breeds, but they can produce more than traditional breeds can on low-quality feeds. In the case of milk production, yields can increase from 1 liter per day for traditional (local) cows to 10 liters in crossbred cows fed only forage grown on the farmer's land. Most farmers who have crossbreeds will also keep traditional cattle, but the crossbred cattle produce best if they are kept in zero-grazing shelters to prevent transmission of ticks and other disease-carrying insects.

There have been some efforts to preserve "pure strains" of indigenous animals, such as the *criolla* cattle of Latin America. Certainly these efforts deserve support and can contribute to the preservation of valuable characteristics, such as heat- or tick-resistance. But if crossbreeding produces animals that have these native traits *and* the higher productivity of exotic breeds, the result can be very beneficial for farmers.

Another way to promote preservation of natural biodiversity is to offer alternatives to the poaching of wildlife and to the overgrazing of communal forests or prairies. Local attempts to improve wildlife or forest management can only work if the fundamental concerns of rural people are addressed first. Farmers can be trained to plant their own pastures and forage trees instead of relying on communal lands or forests for grazing. Many programs require the practice of zero-grazing, in which the animals are confined and the farmers harvest forages they have grown themselves.

The importance of preserving genetic diversity of crops and livestock goes beyond the biological efficiency of the farm ecosystem. In the June 1992 edition of the journal *International Agricultural Development*, the editors predict that the erosion of genetic diversity in agriculture could lead to the greatest catastrophe in human history—widespread starvation in both the Northern and Southern hemispheres. The Food and Agriculture Organization (FAO) has warned that loss of the earth's genetic plant resources is a grave threat to world food security. The livestock world would do well to prevent the crisis that has befallen the plant world and to begin paying serious attention to genetic diversity of livestock breeds before another valuable resource is lost forever.

Soil and Water Protection

Overgrazing and undergrazing are important concerns. Overgrazing is one of the most talked about and least understood concepts in livestock management, and it comes up frequently in arguments against livestock in many places where development programs are being planned. Overgrazing has contributed to desertification in Africa. In the Amazon basin deforestation of tropical rainforests has occurred because of clearing to create pastures. However, it must be recognized that humans, not animals, are largely responsible for this destruction.

If animals are well managed and raised in a balanced farming system, they can enhance the soil and provide incentive to farmers for planting grass and trees to control erosion. Generally, when erosion is controlled, water quality is improved. Furthermore, only 40 percent of the world's land can be cultivated. People who live in places where the land is sloping and rocky cannot grow crops without destroying the soil through wind and water erosion. In these situations, raising livestock is an appropriate alternative to crop farming. *Ruminants*, animals with four stomachs, can digest cellulose that is not edible by humans. They turn scrubby grass or leaves into quality protein for the family or market and produce wool or hair for clothing.

Most people think overgrazing is caused by having too many animals on a particular plot of land. Actually, the number of animals has little or nothing to do with overgrazing. André Voisin (1966), a French scientist, demonstrated that overgrazing relates to the amount of time plants are exposed to animal grazing. In other words, if 200 cows graze on one acre for only one hour, overgrazing may not occur. But one cow constantly grazing on a five-acre pasture will probably overgraze some areas in that pasture.

In his book *Holistic Resource Management*, Allan Savory (1988) writes about the importance of the length of time animals spend on a piece of pasture land as it relates to proper cycling of the plants. This delicate interdependence of plants and animals can be severely affected by mismanagement. Humans have not always understood the importance of looking at the whole picture of agriculture and environment. Savory

defines overgrazing as "any grazing that occurs on leaves that have grown from root, rather than direct sunlight energy." This concept is related to photosynthesis.

Photosynthesis will make food for the plant as long as the plant has plenty of green leaves available to catch the sunlight. If a plant is grazed too severely, not enough leaf area remains to catch the sunlight. In this case, very little photosynthesis will occur. Instead, the plant has to grow back from energy stored from previous photosynthetic activity. If the plant has to do this too often, it could eventually die.

Unless farmers take advantage of ecology, they will be fighting nature at every step. Fighting nature is expensive in many ways. Without animal manure, expensive chemical fertilizer must be applied. Rodale Research Center and others have demonstrated that many chemical fertilizers are harmful to the micro-organisms in soil, destroying valuable life in the soil and leaving the farmer "addicted" to chemical fertilizer. Without birds and predatory insects, plants can be "overgrazed" by insects, and farmers feel forced to use pesticides. These chemicals can kill both the good and bad insects, and farmers end up "addicted" to expensive pesticides as well.

One interesting project in the Dominican Republic uses goats and sheep for weed control on large plantations of sisal and aloe vera. The animals, which do not like these two plants, graze and browse between rows to clean up the grass and weeds that would otherwise need massive application of herbicides.

Examples of environmentally positive practices that can be promoted by development organizations include the following:

- Development of contour *bunds* (ridges) on sloped land using ditches and/or vegetative barriers.
- Emphasis on the importance of manure and urine for fertilization of crops and pastures, while discouraging burning of dried manure for fuel, whenever possible.
- Use of locally adapted and culturally appropriate animals in projects.
- Planting trees, especially nitrogen-fixing trees, for fuel wood, fodder, living fence posts, green manures, and so forth.
- Encouragement of diversity and integration of animals into local farms where appropriate.
- Decreasing animal pressures on overgrazed or overstocked lands while developing cut-and-carry confinement systems.

A "Modern" Family Farm

Lest we give the impression that sustainable farming is only applicable to small-scale subsistence farmers or is a romantic notion of back-to-the-earth ecologists, here is an example from a prosperous farm family in the United States that has made a successful transition to sustainable agriculture.

If anyone has the notion that farming more sustainably is a step backward, they would do well to visit Florence and Dave Minar's farm just outside New Prague, Minnesota. The Minars use farming practices that were labeled by some as foolish and doomed to failure when they first began to implement them almost two decades ago.

The Minars' operation includes approximately 175 registered Holsteins, with a 70-cow milking herd. Feed comes from 230 acres of crops planted in a four-year rotation: corn, then a field pea and barley cover crop for alfalfa, and finally two years of alfalfa. This rotation nurtures both the livestock and the land. No additional nitrogen is purchased. Manure is handled as both liquid and solid, with the liquid custom-applied each spring.

The Minars use no herbicides or insecticides, other than an extremely rare "rescue" application. "Our weed control program emphasizes timely use of the harrow, some rotary hoeing, and one or two cultivations. We don't till in the autumn, and we chisel plow instead of using the moldboard." says Dave. "Our crop yields are generally above the average for the area, and our profitability has been enhanced by the lessened input costs," he added. The herd average has also remained high, at 22,000 pounds of milk.

When asked what sparked their interest in an alternative style of farming, both Dave and Florence point back to the early 1970s and cite a growing discomfort with the conventional, chemically based farming they were practicing. Family health was a part of that concern. "We also wanted to farm more in tune with our values, which include an appreciation for simplicity and being good environmental stewards," says Florence, who feels that an attitude of self-reliance was also important. It helped them overlook some negative feedback they were getting in the early days.

The Minars got serious about changing their farming practices in 1974, when Dave suffered a severe reaction to some agricultural chemicals he was mixing. They began with some trials comparing chemical and nonchemical methods on a portion of the farm. They found little difference in yields and a great deal more personal satisfaction and peace of mind. Dave speaks of going through a "transition period" in which they had to refine some techniques and skills. Most of the shifts occurred relatively quickly, although he feels that some of the more subtle changes may have required as much as 10 years to complete.

Benefits have been numerous and are especially apparent during dry years. "In 1988, our corn held up quite well, looking good right through the drought, and yielded much better than we expected. We credit improved soil conditions and tilth, which have created a great deal more water-holding capacity," explained Dave. "As a result, erosion is no longer a problem."

Farmers and researchers are increasingly searching for low-input and ecologically sound approaches such as those used by the Minars. The advances of modern technology should not be at odds with the best interests of farmers, food consumers, and ecologically minded citizens.

Three

Sustainable Agriculture: Humane and Socially Just

Perhaps the predominant justice issue in the world today is the fact that 1.2 billion people—one-fifth of humanity— exist in shocking conditions of depravation and poverty. The income disparity between rich and poor countries is particularly dramatic. The average per capita income in low-income countries in 1987 was only 6 percent of that in the industrialized part of the world. The following are a few facts regarding human deprivation in low-income countries, according to the *Human Development Report 1990:*

- A sixth of the people in the South go hungry every day.
- Approximately 150 million children under five (one in every three) suffer from serious malnutrition.

■ Approximately 14 million children die each year before reaching their first birthday.

The problem of poverty and hunger in many parts of the world is a monumental challenge. It is impossible to talk of a just and sustainable future without confronting these realities. A corollary challenge to finding ways for people to feed themselves is the problem of the population explosion.

Population change and its effects are among the most important driving forces in the world today. In 1987 the world population passed the 5 billion mark. Lester Brown of the Worldwatch Institute points out that at current birth rates it will double again in 40 years (Brown et al. 1992). A projected 876 million people will be added to the world population during the next decade, enough to completely populate a new India plus a new Ethiopia at their current sizes. This growth will be concentrated in the poorest and most environmentally degraded countries—the very countries that have the least prospect of being able to accommodate it.

Urban populations around the globe are exploding. The United Nations estimates that between 1987 and 2025 the urban population of Africa will grow by 2.75 billion people. The impact of this is all too evident as such large numbers join the ranks of those looking for jobs, food, and income—poverty and hunger on a scale never before imagined. These are people who are driven from rural areas because they think work can be found in the cities. Governments and nongovernmental organizations alike need to help fight this alarming trend by doing everything possible to make life in rural areas a meaningful alternative. This means supporting both programs of family planning and self-sustaining food production. Not doing so makes any programming akin to trying to cure cancer with a band-aid.

Going against nature results in social expenses that hurt everyone. Chemicals pollute water and cause health problems. Factories that produce chemicals pollute the air and add to the greenhouse effect that is warming the earth. In addition, people who work in these factories can be hurt by exposure to the chemicals. In short, fighting with nature is like borrowing from the earth. We would do well to listen to the African proverb that says, "Don't borrow from the earth, because it reimburses itself with interest."

Going *with* the Grain—The Challenge of Rural Development

A large number of the world's poor are rural people who derive a living from agriculture. In many parts of Africa and Asia up to 80 percent of families still engage in farming for their livelihood. Hundreds of millions harvest their own food every day. In most Latin American countries, in spite of rapid urbanization, more than 50 percent of the population is still rural-based. In these countries, meat, milk, and eggs do not come in cardboard or plastic containers, and few people have the luxury of debating the role of livestock. In predominantly rural societies in underdeveloped countries, 98 percent of the large animals, such as dairy cows in certain African nations or water buffalo in Thailand, are raised on small-scale family farms. The land holdings of the families in these societies usually average only one or two acres, sometimes less.

Most of these farmers live at a subsistence level, many of them in poverty. In Burundi, 85 percent of the population was living below the absolute poverty level in 1987; in Cambodia, the figure was 98 percent, and in Peru, 83 percent (State of World Hunger, 1990). The importance of sustainable agriculture to subsistence farmers, including good management and care of animals, is obvious.

As mentioned earlier, many people contend that the raising of livestock, especially meat, is an expensive luxury that our world cannot afford. We would agree that the world cannot afford the level of meat consumption typical of the richest countries. John Robbins (1987) has argued that hundreds of hungry people could be fed on the soybeans and grain consumed by a few cows. This is one way to see the problem of world hunger, and it assumes we are talking about North American beef cattle, which are a small portion of the total cattle population in the world.

Some of the accusations made against livestock are true, particularly as production occurs on modern factory farms. Also, criticism against beef ranching that destroys the tropical rainforest is justified. Nevertheless, when managed humanely on a diversified farm, livestock can make an important contribution to the nutritional status and quality of life of people who live on small-scale farms.

True development focuses on enabling people to care for themselves, their families, their neighbors, and the environment. Neither eating less meat nor producing fewer farm animals and livestock products will, in itself, achieve this goal. The process of holistic change, which we call transformation, has much to do with justice, and transformation involves the people in rich countries fully as much as it does the people of low-income countries.

Equitable development results in both economic and social benefits. In addition to enhancing the self-confidence that results from economic improvements, livestock can be a source of social status. Livestock ownership defines wealth in many cultures. In land reform situations, such as exist in Honduras, livestock are needed to bring land into production once it has been won back by disenfranchised farmers. Without livestock, the land could be taken away from the farmers before they have built up enough capital to bring it all into crop production.

Caring for the Land—A Question of Ethics

In his book *New Roots for Agriculture* Wes Jackson says that early writings of prophets and scholars alike lamented the loss of soils and warned people of the consequences of their wasteful ways. "It seems that we have forever talked about land stewardship and the need for a land ethic, and all the while soil destruction continues, in many places at an accelerated pace. Is it possible that we simply lack enough stretch in our ethical potential to evolve a set of values capable of promoting a sustainable agriculture?" (1980, 13).

Dean Freudenberger raises the question of social justice and human survival when he points out the failure of modern agriculture to sustain the rural community and care for the land. He says the farm crisis ". . . raises doubts about the moral integrity of our culture. It suggests weaknesses in agricultural science, technology, and industry. It suggests a weakness in social understanding about the land and the community of those who cultivate the land" (1990, 37). Freudenberger maintains that agriculture "is the most important activity of every society, and that, unless we reevaluate our relationship to this planet, we will not be able to resolve the agricultural and ecological crisis" (1990, 18).

Concern about the integrity of creation and the development of people leads to a world view with justice as one of its cornerstones. It motivates a search for how to work with people who have traditionally been ignored by both the political process and modern agricultural research—small-scale farmers living in rural, often isolated, areas.

Luckily for the earth, many traditional farmers in the world still know how to care for the soil; more would if they had better opportunities. It is these farmers who protect the wealth of knowledge about biological relationships on the farm, knowledge passed through generations of farmers.

No race can prosper till it learns there is as much dignity in tilling a field as there is in writing a poem.

—BOOKER T. WASHINGTON

Gender Concerns

The role and rights of women is a worldwide social justice issue. In many parts of the world, farm work done by women is not afforded its due respect. It is not qualitatively measured because most of the products of women's labor are consumed in the home or sold at local markets. National and international planners often disregard this dynamic production capacity. Yet, for resource-poor families, home food production is the key to their survival.

Caring for livestock near the home is almost always the responsibility of women and children. When new ways of managing animals or new types of animals are introduced to an area, the women should be included in the training—after all, it is their daily responsibility to care for the animals. Unfortunately, most development programs have not adequately considered the participation of women. Training women requires extreme sensitivity to traditional customs and mores, and female extensionists and trainers may be necessary. Sustainable development is less likely to happen as long as women are excluded or ignored.

Gender sensitivity in planning a livestock development project leads to two major benefits. First, the project has a better chance of succeeding if women are involved in all stages, from planning through implementation. Family nutrition can best be improved when women have the resources and the training to feed their families. Second, when development agencies treat women's work as important, women begin to see themselves as important. This is especially valuable in countries experiencing rapid social change and conflict, where women may feel like chattel even though their countries have laws giving them official rights. Men may also begin to give greater recognition to their wives and daughters when they think of them as productive members of the family.

The transformation to a just and sustainable society, including one with food security, must include respect for women and their many roles.

When the Kechuayamara Foundation started a guinea pig improvement project among the Ayamara people of Bolivia, women and children were the principle participants. As the project became established and demonstrated success, all family members were integrated into the process. This particular project fulfilled the criteria for sustainability on all counts. It was successful because the family members could all participate in culturally accepted ways.

One Woman's Story

A woman from a village in Uganda told the following story:

My name is Jane Bella Magombe. I am married to Kenneth Magombe and we have four children—three boys and one girl.

One day in 1988, Mr. Wesonga found me by the roadside waiting for a vehicle to take me to town. He stopped to greet me, and he asked me why I never participate in some of the activities which take place in the village. I was surprised at his question because I did not know that there was anything worth my attention taking place in the village. Before he left he asked me to stick around the next day and come to a meeting where the people were going to talk about a new project with milk cows.

I went to this meeting and noticed that it was only attended by men. During the meeting I advised those present that since they were planning a project in which the animal was to be zero-grazed, it will inevitably mean that the women would be doing most of the work for the cow, and, therefore, they needed to involve the women and allow them to come to these meetings. I seemed to have made a point because I was fully supported by the outside visitor. From then on, the management of the project moved away from men to women. That was a great landmark in the history of women's leadership in our village, especially in a sensitive area like keeping cows. I became very active in the project and also found a chance to impart some of my knowledge as a social worker. We were moving from the known to the unknown, and we needed to attack the program from all angles.

It was when I attended more of these meetings that I developed the desire to become the proud owner of a cow. My social status compared to the rest of the people in the village was fair on the face of it. But on serious consideration, I found out that I was not any better off than my neighbors. First, I did not even own a local cow, and second, I could not afford daily milk for my family. Third, my daily income was very small indeed, and, fourth, I was undergoing a transitional situation from urban life to village life, and I seemed not to fit in very well.

I went home and told my husband about the project and that I was determined to plant grass and build a shed in readiness. He did not want to let me go ahead. I was determined, so I started planting elephant grass in spite of his protests. At one time, he even threatened to uproot it in preference for eucalyptus trees. The most hurting bit came when I needed to construct a shed for a cow. Here, he just had to give his permission first, because traditionally a wife cannot construct any building in the compound unless the husband has given his permission. Poor me. I did not have that permission.

I started looking for ways and means of involving him so that he could see my point of view. Before long a golden opportunity struck. We had to organize a seminar for the farmers on bookkeeping and record keeping. He knows bookkeeping so I asked him if he could teach these subjects to the farmers. He was very happy to do it, and after the seminar he told me that he would do anything I wanted in preparation for the cow. Within two weeks, the cow shed was ready. He ordered the water trough and the feeding box, and it was only then that I was put on the list of those who qualified to receive a heifer.

December 8, 1990, was the day. Nine heifers accompanied by one bull arrived in our village. There was much rejoicing and dancing. Immediately after I got my heifer I took her home. I consulted with my husband as to what name we should give it. Margaret Thatcher, the former prime minister of Britain, had just resigned from the prime ministership, so I called my cow by that name lest we should quickly forget the Iron Lady's rule. My heifer is called Thatcher.

My husband and I worked hard to make sure that our cow was fed well. We almost followed the seminar notes to the letter.

From that time we started receiving many visitors, almost all coming to see our cow. We did not hesitate to explain to them everything concerning our heifer. Many of them have grown interested and have registered in our project.

Thatcher calved down on 31 March, which was Easter Sunday, bringing forth a healthy bull, 37 kilograms in weight. The assistant extensionist, who is a retired agricultural officer, . . . taught us how to milk the cow. Meanwhile we continued attending seminars on the management of a dairy cow. For the first three months Thatcher gave us an average of 13.5 liters per day. Then we ran out of grass and we had to improvise by giving other types of grasses. The milk yield dropped to 8 liters per day for the next two months. Meanwhile our napier grass was growing.

I have four children, three of them are attending secondary school. During the Easter holidays and other holidays, I did not find a problem in keeping them fully occupied. Right now my eldest son is employed at home to care for and feed the cow at a monthly salary of 6,000 Uganda shillings. The others lend a hand in all ways, like milking the cow, spraying it, and cleaning the shed. Now, nine months later, Thatcher is still giving us milk at an average of 9.5 liters.

We drink 3 liters per day and sell 6.5 liters, which gives us a comfortable 58,500 Uganda Shillings per month. This has enabled me to engage in some other profitable activities. My family is very healthy and happy. In fact, since we started drinking milk, we have not visited the hospital. Our house was at a standstill, but we now have started building it again. School fees are no longer too much of a problem.

Sharing Resources—A Tool for Justice

The continuous process of "passing on the gift," as practiced in many livestock projects, ensures that the project group will always provide capital resources to other disadvantaged community members. The last section of this book deals with this principle in depth.

One of the cornerstones of effective, people-centered development is "sharing and caring." We have observed that a sense of compassion and spirituality is far more obvious among rural farmers with whom we work in many parts of the globe than in the prevailing "materialistic" world view of people in the developed countries. Farmers might need help with economic and material development, but they are often rich in spiritual and social values. Nongovernmental organization staff and volunteers with enough money and time to support the work of development have much to learn about spiritual development from the farmers they support. Those of us who have the privilege of working with rural people often comment that we are renewed and transformed by the relationship.

A Just, Sustainable, and Humane World

A just, sustainable, and humane world might be one in which ". . . all forms of life (plant, animal, human) are respected. The fundamental dignity of all human beings is recognized, and relationships and institutions incorporate such basic values as trust, honesty, self-respect, cooperation, and compassion. The cultural and spiritual integrity of society is preserved and nurtured" (Reintjies et al. 1992, 2).

We live on a planet where much wildlife habitat has been converted to human use. The balance between plants, animals, and humans has changed dramatically as the human population has exploded. Without domesticated animals, the balance might soon be limited to plants and people, with pockets of wilderness and protected wildlife. In addition to the ethical questions, this different balance profoundly affects biological cycles. Such a discussion may belong in the realm of environmental ethics, futuristics, theology, and philosophy; but the urgent work of enabling impoverished rural people already on the earth to meet their own basic needs requires the application of just and humane principles in practical situations with small-scale farmers.

The Humane Treatment of Animals

Because management by humans is the determining factor in the well-being of farm animals, training is an essential part of all livestock development projects. In some cases, the farmers know as much, if not more, than professionals about how to care for animals in their particular setting. But when an alternative technology, such as zero grazing, is being introduced, it is important to ensure that training in animal husbandry is provided before the animals are delivered, and that the training is consistent with animal well-being and health. Training often transforms people who have not had access to formal education into teachers in livestock management. They become natural allies for spreading knowledge throughout the community.

The sustainability of the reciprocal relationship between people and farm animals in a society that has become almost completely urban, such as that of the United States, is a pressing issue. Under modern large-scale confinement systems, the relationships between people and animals change. When too many animals are raised together in a limited space, disease can be catastrophic. Producers rely on antibiotics and treatment for sick animals instead of human care and disease prevention. On the other hand, many small family farmers still operate in labor-

B. Colter

SUSTAINABLE AGRICULTURE: HUMANE AND SOCIALLY JUST

intensive systems, where family members readily give good care and personal attention to their livestock.

Livestock producers in Western Europe, Canada, the United States, and Scandinavia are responding to consumer demands for healthy food, which is seen as an integral part of the animal welfare movement. Sweden has legislated animal welfare codes, and the various ministries of agriculture in Great Britain have published Codes of Recommendations for the welfare of livestock since 1971. The United States has no such official codes, but several groups and individual veterinarians have outlined their requirements for livestock welfare.

The Humane Society of the United States has developed a list of four pillars of humane animal husbandry. These pillars point to the "right" ways to relate to and care for animals: (1) right breeding, (2) right rearing and socialization or bonding to human caretakers, (3) right nutrition, and (4) right environment. Others would add two more pillars: right health care, and right handling and processing.

Right Breeding

Breeds selected should be appropriate to the region and to the system in which the animals are raised. Selection must take into consideration why the livestock are being raised, especially if the goal includes family nutrition or the requirements of the market. Protection of genetic diversity and indigenous breeds is as critical for livestock as it is for crops. Selection to enhance growth and productivity should not undermine natural disease resistance or adaptation to the local environment.

It is wise to provide breeds that are adapted to the local area. Although upgrading the genetics of local breeds is one of the primary goals of many subsistence farmers, the widespread importation of exotic animals should not be encouraged. In many cases good breeds can be found locally. If not, semen or a breeding male can be shipped from a nearby country. Crossbreeding helps animals retain much of their natural disease resistance and adaptation to the local environmental conditions.

Nongovernmental organizations such as Heifer Project International work to protect diversity by promoting the use of indigenous breeds in many parts of the world—including guinea pigs in Peru, alpacas in Bolivia, Ndama cattle in West Africa, Bengali goats in India, yaks in China, and indigenous water buffalo in Thailand and the Philippines.

Right Rearing

Strong bonds have linked humans and livestock for centuries. Many livestock systems in the industrialized world have become so mechanized that human contact is minimal. But two-thirds of the world's farmers do not have ready access to fossil fuels, chemicals, or mechanization. Small-scale development projects typically deal with these nonmechanized family farms, where human contact is a natural part of livestock care. In cases where the traditional approach to caring for livestock does not ensure clean water, adequate feed, and protection from weather, the project should provide training on the importance of these factors. In projects that use the "passing on" requirement, the local community organization itself monitors implementation of quality care because of the vested interest members have in quality control and guaranteeing the best possible stock for the "pass on" to future beneficiaries.

Right Nutrition

Healthy animals require good nutrition. In low-income areas where veterinary services are critically lacking, proper nutrition can prevent 80 percent or more of the most common health problems of livestock. Adequate amounts and quality of both feed and water are vitally important. Keeping water available for livestock is not always a natural practice for people, so livestock extensionists need to remind farmers of this important element.

Most of the training given to the farmers through successful development projects involves supplementing their existing knowledge of livestock nutrition. Farmers examine the basics of livestock nutritional needs and how ruminants differ from monogastric animals. They plant grasses and nitrogen-fixing trees to supplement the natural pastures (and to provide erosion control as well). In many cases farmers learn about industrial or agricultural wastes and byproducts available in their areas. Ruminants on small farms (cows, goats, sheep) are seldom fed grains, and hogs and chickens are often fed residues or feeds that are not fit for human consumption, or at least do not compete with family nutritional needs. Farmers realize the trade-offs and usually request animals that will not compete with their children for food.

Cattle feed supplements include bran, cottonseed, or oilseed cake rather than grain. If grain is fed, it is the choice of the farmer, based on the farmer's judgment that value is added to the grain in feeding it to hogs or chickens. For example, in Zimbabwe farmers can get two to three times the return on surplus grain if it is fed to hogs rather than sold directly. This practice also makes marketing the grain simpler. Transportation to market is expensive if the quantity is too large to be carried on a person's head or back. In Zimbabwe, however, the farmers must weigh the benefits against the risk of droughts, when the grain-fed pigs would probably have to be sold because of grain shortages.

In any case, all livestock development projects can only work if the farmers can feed the livestock adequately. Feed sources must be readily and reliably available, and pastures or trees must be planted and producing before animals are brought to the farms.

Right Environment

In addition to giving shelter from harsh environments, housing should provide access to exercise, good ventilation, and comfortable bedding. In tropical environments, protection from the sun is an important consideration. Sound manure and urine removal is also important in humane housing for confined animals.

Animals are kept in a wide variety of housing, depending on the traditions in the area. For example, the Kenya National Dairy Development Project recommends a standard zero-grazing unit for individual farmers with an average of three to five dairy cows. The primary purposes of this popular unit are to allow for good nutrition for the livestock, good ecology for the farm, and good health for the animals. In addition to making feed and clean water constantly available, the system protects the animals against ticks and allows for close observation by the caretaker. Manure and urine are easily collected for use as fertilizer or biogas, and each cow has her own covered resting area with clean soil for bedding, as well as access to an outside exercise area.

In traditional rural cultures, most animals are cared for by a particular family member who takes the animal to pasture and water and brings it home at night to sleep in a part of the house or in another enclosure.

When the traditional methods of livestock care do not meet humane standards, training and other assistance may be needed. It should be kept in mind, however, that in some climates and environments, appropriate protection and housing may be available naturally and do not need to be constructed by humans.

Right Health Care

Prevention (right nutrition, vaccination, dips or sprays against parasites, etc.) often makes other veterinary drug use unnecessary, except in emergencies. Prevention of health problems is much better and less costly than treatment. Veterinarians, whether on the staff of the development agency or working in private or government practice, are an important resource in all project areas. Infectious diseases can be very dangerous to the animals. Often available vaccines can prevent them. But it is always wise to talk with animal health care personnel about the local situation.

Many projects have successfully included training of local "para-veterinarians" and extension workers to manage routine health care. Nongovernmental organizations are usually more able and trusted by the communities to distribute veterinary supplies and services. However, cooperation with local veterinarian authorities is always counseled.

Locally trained animal health workers often combine the best of modern skills with time-honored local practices. Projects in several countries emphasize the importance of recording traditional herbal remedies for livestock. One project in the Philippines, the Community Animal Health Volunteers program, relies largely on indigenous remedies and promotes preventive practices. Another in Cameroon helps to preserve medicinal plants and recipes used in traditional treatments.

Right Handling and Processing

Good husbandry includes handling, such as castration of males unsuitable for breeding, and disbudding (removal of horns before they form) when appropriate. These procedures are done on young animals so that discomfort is minimized.

In projects involving farmers with limited resources, livestock are often raised for a variety of reasons more important than slaughter for meat. Many poor people only slaughter valuable animals for special occasions or when the animals are old or sick. Otherwise, cows are raised primarily for milk or for offspring to sell; water buffalo, donkeys, and oxen for draft power; hogs for income from sale of offspring; and chickens for eggs. Goats are raised for both milk and meat, though people in many cultures around the world raise goats primarily for meat. For those occasions when a farm family might serve meat in the home, they may buy small quantities at the local market. When farm families need cash they may sell offspring (such as goats, pigs, or male calves) for income. These animals will most likely eventually be slaughtered for meat. For those who would like to eliminate all slaughtering, no standards meet their criteria; but many traditional methods make slaughter as quick and painless as possible.

Directions for the Future

Many organizations continually develop criteria that can serve as standards for the care and management of animals. These guidelines should include provisions for housing, feeding, transporting, working, rearing of offspring, restraining, marketing, and slaughtering. Professionals, volunteers, and producers need to consider the well-being of livestock in terms of ethics, economics, and ecology. The health of people, the land, crops, and livestock cannot be separated from each

other in integrated farming systems. Keepers of livestock should know more about how human contact and attention affect growth and development in animals, how dirty water carries disease, how housing affects health, and how a balanced diet contributes to overall health.

For these reasons, education is one of the cornerstones of any livestock project. When farmers are trained in animal husbandry, they are able to pass on this gift of knowledge. Giving of oneself to others is an appropriate vehicle for building confidence and dignity.

Part

Two

The Process Matters

Holistic transformation involves a change in the whole person: physically, mentally, socially, and spiritually. The role of the outsider, whether an organization or an individual, is to facilitate a process of change. This can be a long-term process. The focus is on learning and growth rather than on a material product. In one sense, the process is the product.

What type of process will lead to transformation and the genuine development of limited-resource families? We suggest that this process must have certain characteristics. It must be *participatory, educational, sustainable, holistic, and grounded in the community.*

Some of what follows may appear to be a linear, step-by-step process. We do not wish to leave that impression. Ideas are presented in a logical manner for clarity rather than to suggest that there is a series of clean, neatly distinguished steps that can be followed as a recipe in the work of human development. Indeed, these concepts and principles constitute a set of values and an approach to people-centered change. Transformation is more of an art than a science. It is as much a commitment to people as a method.

The livestock development project is a part of a larger whole. It is necessary to look at the family and community holistically. Part of the art is in asking the right questions. Will the proposed project generate income to help send children to school or require them to herd animals? How will providing women with training and livestock affect their relationship to their families? Dignity and pride come from providing for one's own family. Being able to assist a neighbor with milk for a sick child may justify an activity which, when looked at in strictly economic terms, might not bring a very high return.

Environmental Impact

The first section of this book looked at the characteristics of sustainable agriculture. As we begin to explore the application of these principles in small-scale livestock development projects, we do so with a serious consideration for the environment. A wide divergence of opinion exists regarding farming methods—should they be organic or conventional? What about integrated pest management, use of inorganic fertilizer, and fossil fuels?

The environment, the quality of human food and groundwater, soil and fertility loss, and deforestation are now widespread concerns and are among the most serious problems facing the international development community. This is a concern not only to people in high-income countries, but also to the millions of people in villages and rural areas of low-income countries. Some of the most direct action being taken to combat environmental degradation is being done by people living in communities of low-income countries.

We affirm the importance of the process of human transformation and believe that the planning of livestock projects must be done with careful attention to various processes and expected outcomes. One of the most important outcomes has to do with the natural resources all projects deal with or manage. Just as we care about people, so we must care about the earth and its other creatures.

Four

People Will Say "We Did It Ourselves"

Alhaji Amadu Buba is a tall and distinguished-looking Fulani elder. His beautiful copper complexion is set off by a sky-blue flowing robe and an intricately embroidered cap. Alhaji is president of the Council on Ethno-veterinarian Medicine in Cameroon. When asked who made the decision to start this project, he immediately answered, "HPI come join we."

Understanding pidgin English is not always easy, because of double negatives. Phrases join opposites, as in "I go come." Many truly unique uses of words, which may sound familiar to an English speaker, mean something quite different. For example, "fine chop" is good food. However, "HPI come join we" needs no translation. The sentiment expressed here is that Heifer Project International joined the Fulani in their effort to develop themselves. Other projects in Cameroon focus on helping people with this "join we" approach, including projects with rabbits, dairy cattle, multipurpose cattle, oxen, agroforestry, and ethno-veterinarian (traditional) medicine.

One of the cornerstones of development work is self-reliance. This begins with people choosing the problems or needs they want to address.

The Participation Principle

Participation by the community is now a widely accepted principle of development. No one opposes it, but it is applied in many different ways. Sometimes it simply means asking people what they think. At other times it may involve participation through physical labor. Purists say this principle means giving people complete control over planning and fund management.

Those who are committed to participation promote consciousness raising and try to help marginalized people claim their rights and regain self-esteem. Staff and volunteers of development agencies have thou-

sands of stories to tell of working together for the common good at the community level. To listen to these stories from the poor themselves can be tremendously uplifting. One of the privileges of working in the development profession as a promoter of participation is to see people's lives changed and improved through small-scale projects. The content of the project may be livestock improvement or water systems or community health committees. How it is done is fully as important as what is done. To see participatory development is to see the powerless girding themselves with strength (Aaker 1993).

Outsiders should avoid the simplistic approach of only asking for opinions or presenting the group with a predetermined package as the solution to their problems. Too often in the past the "specialist" has defined the solution and then asked the people to help implement the idea. While the latter might result in a successful production activity, it does not build people's ability to control their own development.

Community people should influence key decisions and be involved during all stages of the project. This includes determining what needs or problems people want to work on, forming a group, deciding on the type of action to take, obtaining the needed resources, and implementing activities. Certainly community people have a stake in the outcome. In fact, they are the owners of the project. This type of participation can be referred to as "responsible ownership." The project and the process belong to the people. They feel responsible and hold each other accountable.

Patrick Breslin of the InterAmerican Foundation wrote about people involved in grassroots participatory development in these terms:

> And what they had in common, every one of them, was dignity. I became convinced that no matter how abysmal the economic level, how desperate the need for assistance, there is no point at which the dignity of the person is not more important than the aid itself (1987, xii).

Participation is the underlying concept that should permeate all project work. The role of the outsider, technician, or leader should always be that of catalyst, teacher, and facilitator. The facilitator's aim is to enable people to develop the skills, motivation, and resources to do it themselves. The idea is well conveyed by the proverb, people will say "we did it ourselves."

Help That Empowers

The diagram on the following page shows how requests for help should be responded to in a way that empowers people rather than creates dependency. While the world witnesses a continuing succession of human disasters that compel response, we strongly feel that long-term solutions will come only out of people-centered development processes. It is important for caring people to respond with food aid in

Flow Chart of Two Types of Help

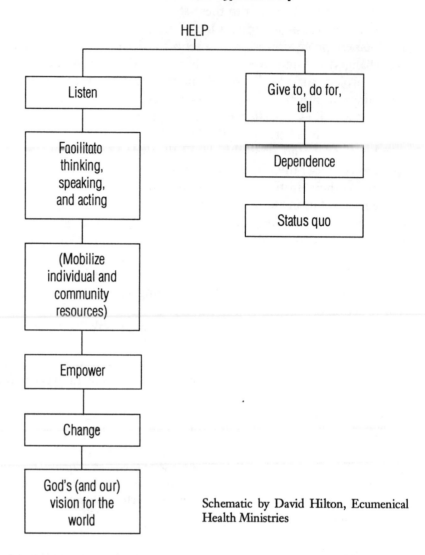

HELP

Listen

Give to, do for, tell

Facilitate thinking, speaking, and acting

Dependence

(Mobilize individual and community resources)

Status quo

Empower

Change

God's (and our) vision for the world

Schematic by David Hilton, Ecumenical Health Ministries

the face of starvation, but the only sustainable answer is to help people feed themselves.

"How To" Ideas

Participation is best achieved through a group process. Through groups with elected and accountable leaders, people begin to exercise control and ownership. By pooling resources and assisting each other, the poor are able to have a significant impact on their own futures. Such cooperation can start with small community groups of 10 to 20 families, but development programs will eventually have a more sustainable impact if larger regional or national organizations and "people's organizations" come into being.

It is also important to involve and educate a broad range of community members, including people of different ages, gender, and social status. Different members of the family or community will most likely have different priorities and views regarding such resources as available labor and land.

It is common, for example, for women to place higher priority than men on food production because of their heavy involvement in both producing food and feeding their children. Men often place lower value on women's labor than on their own. In West Africa men are extremely interested in obtaining draft animals to cultivate more land. The result, however, has often been that women, who already put in long workdays, spend even more time planting, weeding, and harvesting.

Development projects done in partnership between locally organized community groups and nongovernmental development organizations should be carried out under a cooperative agreement. The ideal calls for equal opportunity for participation of men, women, and youth. Although this ideal is seldom perfectly realized, equitable partnership is a worthy goal to strive for.

Needs and Opportunities

Before initiating any livestock development program, it is important to determine what needs people have and what resources are available or needed to meet these needs. One of the first parts of any request for assistance should be the village people's own statement of their needs for a livestock project. This is how the group begins to gain a sense of ownership, and it improves their chances of successfully confronting future challenges and disappointments.

A group may not be fully aware of their needs or the available resources, and so techniques that promote group and team building may be needed. Because the whole process focuses on people initiating change in themselves and their communities, education and participation are, in fact, basic.

Elizabeth Mukong, a large-boned, tall, and beautiful woman with farmer's hands attended the WiLD (Women in Livestock Development) conference in the United States and returned to her village in Cameroon determined to start a WiLD project. Program staff from HPI joined her in studying the concerns of women in her village and found that many women did indeed want to have "their own project." The women of Babanki-Tungo village wanted to start with rabbits.

When an evaluator later asked Elizabeth about the project, she said, "The rabbit project is doing fine; we have plenty to chop and have sold many." The women have already set up a bank account for their savings, and, as is typical of people in so many villages, they want to expand the scope of their project. Recently the "WiLD Action Group" started a tree nursery. Elizabeth notes that firewood and the environment are concerns for all the women in the group. They are determined to do something about it.

As president of the group, she readily points out to the evaluators that the group is now ready for oxen. "We've already been given the land and want to produce food for our families, as well as for the rabbits and poultry. But without oxen we don't have enough power and time to grow these crops, plus food for our families."

The program staff in Cameroon talked about their experience with these women, saying, "It's hard to keep up with these groups, let alone lead them. Much of the last three years the program has focused on helping farmers to get organized into small, self-help groups and to develop the leadership skills of their chosen leaders. The groups are now joining together in clubs or 'pre-cooperatives.' Some eventually hope to evolve into farmers cooperatives."

Some Limitations of the Participatory Program

While collective action and solidarity give the participatory approach its strength, some very common human characteristics are the cause of many failures. Egoism, apathy, greed, and the desire to get ahead at the cost of others are all negative influences and obstacles to participatory development. Also, women in patriarchal societies have immense difficulty speaking for themselves. These characteristics are not exclusive among the poor, of course (Aaker 1993).

Unrealistic expectations can also be a problem. It is not only important to determine the presence of a high level of interest in a livestock-related activity; it is also crucial to know that basic resources exist, or can be provided, to meet the need in a sustainable manner. Often poultry projects are started because of a high demand for meat and eggs. Poultry are very adaptable to different climates and geographical settings. However, experience with various projects, especially those in which poultry are raised in confinement, shows relatively high rates of failure. This is primarily because supplies of feed, chicks, and vaccine—inputs that are most likely not produced on the farm—tend to be unreliable and costly. As a result, price fluctuations of production costs and salable produce are often a major problem. It may be better to try a project with low external input costs, even if this results in lower production (e.g., upgraded, free-range backyard poultry). This type of production activity is likely to be more sustainable in the long run.

Simply knowing about the possibility of a better life through the help of an outside agency can begin a process of dreaming and thinking on the part of communities. But certain risks need to be addressed very quickly. Sometimes people in poverty situations are grasping for straws, looking for any help they can get. The danger is that they will begin to think that outsiders can solve all their problems.

Raised consciousness and new knowledge can sometimes lead to frustration. For example, women may consider the potential nutritional benefits of a production project to be very important. Women put great importance on nutrition, especially when they can no longer breast-feed their children, and they need to acquire basic nutritional knowledge. However, a mother may become very frustrated if she learns about good nutrition and feels motivated to provide good meals but does not have the means to accomplish this. Patience is counseled. It may take up to a year to learn how to raise dairy goats, and the benefits of having milk for children may not come into play until the second generation of goat offspring come into production.

Determining Needs

Personnel of development organizations use a number of methods for determining needs and opportunities. These include rapid rural appraisal, participatory evaluation, community meetings, the use of planning and screening guidelines, and the use of established criteria for environmentally sound, small-scale livestock projects. Experience shows that the staff of nongovernmental organizations can benefit from training in the techniques and approaches of needs assessment and planning and evaluation. Resources should be allocated for such training.

In considering needs it is also important to think holistically. Even though there are situations that call for immediate short-term food relief, in most instances such efforts should be carried on simultaneously with efforts to begin sustainable, participatory development.

Beyond nutrition and income, people have many other needs. Some of these are security, belonging, meaningful relationships, self-esteem, dignity, and fulfillment. As noted previously, animals can fulfill a variety of needs and roles. Often the strongest motivation for a project is something other than food or income.

The Fulani, Maasai, and other pastoralists define their well-being and social relationships in terms of animal ownership and exchange. Even if they could earn their living more easily by growing maize and beans, they feel they would loose their cultural identity by turning to crop agriculture. Thus, it is important to define needs appropriately, in terms of the local culture and values.

Another factor to consider is that not everyone is equally needy even within the context of a particular rural community. Often the wealthier members of a community come forward first if there is a prospect of outside assistance. Agricultural extension efforts might focus attention on these "change agents" or "progressive farmers," thereby contributing to the gap between the richer and poorer community members. It is easy to focus on the more "progressive" elements because such farmers are usually ready to take advice. They have more resources to implement and experiment with new technology and tend to be in leadership positions.

A Matter of Equality

Development assistance, particularly from a humanitarian or church-related agency, should have a bias toward working with the poor. This is the group with the greatest need, both in terms of improving living conditions and of requiring outside assistance to do so. In a real sense, the challenge and opportunity are often greatest within the poorest segment—the widows, the large single-parent family, the landless, minorities, women, children and youth. These are people who have been neglected by the normal social and economic structures of society.

Contrary to common assertions, such people are often open to learning and to making significant change. What is perceived as resistance to change may in fact be a cautious attitude forced on them by severe resource constraints and living conditions. They are left with very little room for error. In fact, the poorest of the poor have a very small margin for risk. One crop failure may put them at a starvation level unless outside assistance is provided. However, with some assistance the poorest may well be the most ready to make a change.

It is debatable if an outside development agency can work with the poorest of the poor in livestock development. Rather, the focus of attention is usually on those who have the minimal level of resources necessary to make animal production feasible. For example, a project that provides dairy cattle must focus on truly needy families with enough land to keep a cow and a calf. In the case of a goat, the minimum necessary start-up resources might be the family's available labor and permission to cut grass along the road.

Need should always be a criteria for developmental assistance. The criteria for selecting group members and project beneficiaries must be clear in order to avoid conflict and charges of favoritism.

Five

Training for Transformation

*People hunger first for bread
and second for truth.*

—DAN WEST, FOUNDER OF HEIFER PROJECT INTERNATIONAL

Olgilai is a small, densely populated village on the slopes of Mt. Meru, near Arusha, Tanzania. Most of the 450 families who live on the side of this mountain farm less than three acres of land, and many farm less than one acre. Land is at a premium, and farmers have been slowly encroaching into the nearby forest reserve. They commonly farm slopes at angles of 30 degrees or steeper.

Tradition runs deep among the Warusha of Olgilai, as it does in many tribes and villages. Colorful beads and cloth, stretched ear lobes, shaven heads, and round, thatched huts scattered in family clusters on the hillside are all visible signs of a way of life and farming that seems to have changed little over the last century. What, when, and how to plant, and who plants, weeds, and harvests, are all established by tradition. Almost all families still cook on an open hearth made from three stones, using firewood, a natural resource that is becoming more and more difficult to obtain.

It is therefore surprising, perhaps miraculous, that, due largely to the work of one man with some outside assistance, a quiet "green revolution" has started in Olgilai. A situation that seemed hopeless a few years ago and that was a source of constant conflict between farmers and foresters is being turned into a model conservation program. While at one point the only way to "save" Kivesi Hill seemed to be to forcibly remove the people, now people are being seen as part of the solution.

Loti Sareyo Sandilen, a father of six and a farmer, is a vivacious man of about 40 who laughs readily and robustly. He is also a visionary and a committed Christian who seems to have planted a seed that germinated into a solution to Olgilai's problems: deteriorating soils, lower production, lack of firewood, low income, and malnutrition. He cares deeply about his family, his village, and the land.

According to Loti, it all started in 1988 when he and his wife, Lightness, planted sweet potatoes in their field on Kivesi Hill. From a distance this small hill is a beautiful patchwork of fields with a variety of crops, such as corn, beans, potatoes, and cow peas, crowned by the small remaining forest reserve. A closer look shows that soil erosion on the 30- to 40-degree slope has caused dramatic drops in crop yield, particularly toward the top of the hill. Scarlike gullies run down the hill between stunted rows of corn. Farmers have not been allowed to plant such annual crops as bananas or coffee because this would amount to giving them ownership of forest reserve land.

A few days after Loti and Lightness planted potatoes, it rained heavily all night. They woke in the morning to find all their potatoes washed down to the bottom of the field. Loti realized that with the potatoes had gone tons of his best soil. He knew that unless he did something, this field would soon be useless. Having seen people elsewhere make contour *bunds* (ridges) and terraces, he set to work making bunds and drainage ditches across the slope and planting napier (elephant) grass to hold the soil. It was not too difficult for the soil conservation extensionists to teach Loti the right techniques, given his high motivation at that time.

Like those who mocked Noah's building of the ark, people mocked Loti and Lightness for all their effort, which would surely come to naught. The traditional system had worked for many years, the neighbors reasoned, and the erosion and declining fertility were matters beyond a farmer's control.

The Sandilen family persisted, and soon their efforts came to the attention of the Soil Conservation and Agroforestry Project of Anumeru (SCAPA). This new program, aimed at reversing the dramatic soil erosion on the slopes of Mt. Meru, saw Loti as a natural leader with a strong interest in soil conservation. SCAPA trained Loti in laying out contours, planting trees, composting, and other conservation practices. Loti quickly used this knowledge on his own farm, correcting some mistakes he had made earlier. He also began helping a few farmers who became interested in the program after seeing the healthy crops the Sandilens produced and hearing that the government now sanctioned this effort.

A patch of beans grows between contour rows of lucaena bushes and napier grass on the Sandilen farm. Lucaena is a nitrogen-fixing tree that is an excellent source of high-quality forage and firewood. Napier grass is the main staple for cut-and-carry feeding of dairy cattle. The contrast between the Sandilens' farm and their neighbors' is dramatic. At the end of the dry season beautiful rows of forage and corn can be seen in the Sandilens' fields, while the neighbors' crops are stunted. Much of the Sandilens' success is the result of the widespread use of compost and the better absorption of water because of farming on the contour.

Because of his leadership and positive attitude toward service, Loti was nominated by SCAPA to become a "farmer motivator" in a project supported by HPI in Tanzania. Farmer motivators are volunteer extension workers who develop model integrated farms that include zero-grazing dairy units and forage-bearing trees. They then teach others in their village to do the same.

The Sandilens received a pregnant heifer, which prospered well on the napier grass and lucaena leaves. With the promise of heifers, many other farmers are now eager to plant forage trees and napier grass and implement soil conservation practices, all of which, along with training, are prerequisites to receiving a project heifer. In fact, the SCAPA professionals can't keep up with the demand for measuring contours and providing tree seedlings. Loti and another trained farmer motivator have stepped into the gap to lay out the contours and teach their neighbors what needs to be done. Children at the village school have started a tree nursery to supply seedlings. The government has abandoned its plan to move the people to save the land and is now encouraging the planting of trees and other perennial crops.

Loti has already cast his eyes on another nearby hill. "That one is next," he says with a smile.

The Learning Approach to Development

Training and education have become constant components of the work of development agencies. Sometimes the term *training* is used when referring to skills taught to the project participants, and the term *education* refers to knowledge imparted to the constituency about global and poverty issues. Whatever term is used, the central issue is that of learning.

Participants in livestock projects need to learn specific skills related to care and management of their animals as well as increased knowledge of management and leadership. Everyone needs to learn about caring for creation and what we can do to help transform our world into a more equitable and just place for all.

Development involves more than the introduction of a different animal species or new technologies. Education for groups of all ages and genders, together with appropriate resources, are key components in the development process. Change in people entails learning new ideas and values. In a real sense all aspects of a development process can be regarded as educational. We not only learn from a seminar or discussion. We also learn from our own work experiences and human relationships.

Learning from Experience

The much-used phrase "learning by doing" is not quite accurate; another element needs to be added, because sometimes people do not seem to learn from experience. Action needs to be followed by *reflection* on what has been done, and by *practice*. Reflection is a kind of evaluation and is often most successfully done with other people. Feedback from others helps us learn and think about the meaning of our actions and whether what we are doing is productive and beneficial.

When teaching technical as well as leadership skills and knowledge, we can think in terms of this kind of cycle:

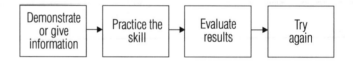

Appropriate and Practical Training

Almost all successful development projects supported by development agencies contain a strong component of training, with the emphasis on appropriate content and methods. The training usually reinforces two important areas of the project: organization and production. Thus, the effort is aimed at consciousness raising and empowerment as well as at increasing technical capacities to improve living conditions.

Perhaps the most effective approach to training that has been developed in the last several decades is that of training community-level promoters. This has been a particularly strong feature of primary health care programs in communities that lack medical facilities and attention. The same principle has been applied to the training of animal health care workers and leaders of grassroots projects.

Across the world we see many attempts at training in development programs, but evaluation of these efforts is notably lacking. Admittedly, it is difficult to know if training is effective or not. Only after the test of time can change and the application of new skills and knowledge be observed. The proof of effective education are changes in both behavior and attitudes.

Consciousness Raising

In Latin America, the 1970s saw the growth of the consciousness raising school of thought as a key, though controversial, method of working with grassroots movements. *Concientizacion*, as advocated by Paulo Freire, implies a liberating education—an educational process whereby people discover their own reality and act upon newfound awareness of their situation (Aaker 1993).

Early in his career, Freire became involved in church programs as a way of addressing social problems and poverty in northeast Brazil. Before long he became aware of the contradictions involved in charity work, with the benevolent middle class doing things *for* the poor but not *with* them. He turned to adult literacy, a critical issue at a time when illiterate people were not allowed to vote. Freire became famous for his efforts to incorporate critical thinking into literacy teaching as a way to raise consciousness in people about what Freire called their "own reality," by which he meant their political and social situation.

Getting Started

Roland Bunch of World Neighbors (1982) stresses the importance of early success in the process of working with people as individuals or communities. If the first experience with a program or a technology is negative, it is hard to overcome this in subsequent efforts.

Each activity should be regarded as an educational opportunity, deliberately pursued for this purpose. The provision of an animal or any other input, although important in itself, is most significant in terms of how it can affect the whole family and community. In Tanzania, the prospect of receiving a quality dairy heifer is the primary motivation for people to begin to learn and practice soil conservation and agroforestry. The dairy project is also the main factor leading to the formation of village dairy associations, which often engage in other development activities. The provision of 20 dairy heifers for a village group not only changes their farming system, but also their outlook on life.

Trust is an important foundation upon which to build successful teaching encounters. Sometimes opportunities to create such trust occur unexpectedly, as the following anecdote illustrates:

> The day after a U.S. Peace Corps volunteer arrived at his new post as the animal husbandry officer, he was aroused early by farmers calling for help with a sick and apparently dying cow. Based on clinical signs, but without a clear diagnosis and with no experience, he decided to treat the animal

with a dose of antibiotics. The cow recuperated. The volunteer was off to a great start. It could easily have been the opposite (DeVries 1992, 2).

The above example is not meant to suggest that an inexperienced volunteer should take a chance with a farmer's dying cow, or that antibiotics is the preferred treatment. Obviously, an outsider should not guess at solutions to every problem that presents itself. Rather, the volunteer related this as an example of how important it is to build trust as part of a teaching relationship, albeit in this case more by good luck than by skill.

Phases of Training

Training in livestock development programs has four distinct phases: (1) planning, (2) preparation, (3) implementation, and (4) continuation.

Planning is a time for the volunteer to get to know the group or organization, and vice versa. During this phase the group might need to improve its organizational, goal setting, and planning capabilities. They may need help in doing a simple feasibility study. At this time the group develops and writes its plan and proposal.

In the preparation phase the organization or community has already located resources for a project, but may not yet have received animals or funds. Some beginning support may be needed for training or a pilot project. Quite probably the group needs to improve its accountability skills, work out the "pass on the gift" method, learn about animal housing, establish feed supplies for the animal, and develop a plan for breeding and marketing. During this phase the group is gaining strength and confidence, and they need to work on group cohesiveness and communication.

The implementation stage starts when the project organization and the participants receive livestock and other resources. All of the skills and knowledge the participants began to learn in the previous stages need to be reinforced during this phase. Environmental improvement and production should be stressed.

When the continuation phase is reached, everybody involved in the project has considerable experience. The project may be going through evaluation and replanning. Ideas for new opportunities and development should be explored. Weaknesses in the project and in the organization should be identified for eventual corrections, some of them through education.

These four phases correspond somewhat to Korten's three stages of learning in his theory of the "learning approach to development," shown graphically in Figure 1.

In Korten's "initiation stage" people learn effectiveness. They learn how to do it (whatever "it" is). In livestock projects people learn the most effective way to produce and manage their animals and other natural resources. During this period people make mistakes and learn from them. They may need extra time and outside resources to get started. This "learning to be effective" stage roughly parallels the first two phases mentioned above, planning and preparation.

Next, the project participants need to learn efficiency. They learn to do more of what they do well, and they do it more efficiently. Production is accomplished with less effort. It is especially important at this stage for the participants to begin to learn how to get along without external resources. They should obtain better results with the same or fewer resources. This parallels the implementation phase.

The last of Korten's stages is expansion. At this time it should be possible to increase the number of people who benefit from the project. Learning how to expand benefits works particularly well in the case of livestock projects, as the "pass on the gift" system takes hold. By this time communities and groups should not need many external resources. Nevertheless, continued support from extensionists and field personnel for training and evaluation may be justified. There should be more

Learning to be effective	Learning to be efficient	Learning to expand
Planning and preparation	Implementation	Continuation

emphasis on networking and multiplying the impact through sharing of knowledge, skills, livestock, and grass-roots leadership.

Training Content

Training covers many topics. Each phase may require different content, and the training will depend on who the trainees are and what their needs are. Curriculum and materials are needed in the following areas of each phase:

- Planning phase—Needs assessment, planning, goal setting, project design, group formation, feasibility studies, "passing on the gift" system.
- Preparation phase—Animal housing, health care, record keeping, nutrition, pasture and forage improvement.
- Implementation phase—Breeding and reproduction, nutrition, health and veterinary care, accountability, reporting, marketing, leadership, agroforestry, group development.
- Continuation phase—Networking, environmental impact, passing on the gift, evaluation, marketing, organization.

Making It Practical

Many participants have only a few years of formal schooling. Thus, all training should be as participatory, practical, visual, and dynamic as possible. In the Santa Cruz area of Bolivia, for example, participants themselves prepare flip charts using drawings to visually show animal care and management practices. Participants take short field trips to visit families already practicing a certain animal skill. The farmers themselves participate in the teaching.

Providing the opportunity for limited-resource farmers to receive practical training in environmentally sound farming systems is a good foundation for any sustainable livestock program. But training should include more than learning knowledge and new skills in livestock management and production. The teacher needs to structure a learning environment that motivates the participants to internalize a positive attitude toward putting newly learned skills to use.

The bulk of the training should be hands-on and done on the farm or a convenient nearby site. Training at institutions is often expensive and puts too much emphasis on theory and classroom learning. The average farmer sometimes does not easily understand professionals. Often the institutions such professionals come from have resources such as tractors, electricity, and laborers, making what they practice and advise seem less relevant to the small producer who lacks these resources back home.

For a village-based project to succeed, the group must develop a sense of trust and cooperation, both among themselves and with the nongovernmental agency or government agency with which they work. Training in careful record keeping and accountability reinforces this trust and honesty.

Inclusion of Women

It is important to include women and youth in training activities. Too often training programs teach only men, who, supposedly, go home and tell their wives and children what they have learned. This approach just does not work. Cultural or social barriers often inhibit men from effectively teaching women.

Pusu Women's Group in Kenya was established in 1980 with 30 members—20 women and 10 men. Of these, only 11 women and 2 men were active. Since that time the group has grown to a membership of 68 active members, of which 50 are women. *Pusu* is a Luo word that means "to sprint at once and go ahead." The project has already demonstrated the feasibility of keeping good dairy cattle in this area of Kenya by using the zero-grazing system. The project was successful because of three key ingredients: (1) the acquisition of new skills and knowledge, (2) the availability of resources (the dairy heifer and feed grown on the participants' small farms), and (3) motivation.

Before she became a member of Pusu, Deborah Ogada planted subsistence crops to provide food for her family. She had no excess crops for sale, and her income was very minimal, as was her knowledge of farming. Since joining the group she has become a full-time farmer, and hard work has paid off. She now feeds her family better and sells excess milk for income. Recently, her napier grass and dairy cows received second-place awards in the South Nyanze district farm competition. She was given a wheelbarrow. Like the other women in the Pusu group with dairy cows, Deborah Ogada is largely independent and self-supporting instead of relying entirely on her husband to provide almost everything.

Selecting Trainees

One effective and widely used approach for helping farmers of both sexes continue to learn is the use of village-based, paraprofessionals as *animators*. Other commonly used terms are *promoter, farmer-motivator,* and *lead farmer*. These community leaders, who usually work as volunteers, bridge the gap between professionals and the other farmers on a daily "in-village" basis.

Project groups should elect two persons to receive training in animal production and management practices. This will ensure a community-based multiplier effect. Two trained leaders with knowledge, skills, and motivation can better meet future challenges and frustrations. This also reduces the risk of one of the leaders dropping out or leaving the community.

These paraprofessionals should teach primarily by doing and demonstrating. They should practice what they preach on their own model farm. In fact, this should be a prerequisite for being selected for this role. Although these people work for the benefit of the community, their reward comes largely from the status and recognition received and the benefits of the practices they learn and implement on their own farms. They can be reimbursed for expenses or given some "incentives," but paying them a salary is usually not a sustainable practice. Eventually the farmers must absorb all costs.

In Cameroon, lead farmers are rewarded for their efforts on behalf of their group through work done by members on the leaders' farms. In Tanzania, trained farmer-motivators are the first in their village to obtain tree seedlings and a pregnant heifer. Thus, they have a chance to be the first to benefit and enjoy an initial market advantage. In Bolivia, a plan was adopted in which peasant field technicians obtained in-kind loans of three to five heifers to establish foundation dairy herds.

In all cases the leaders are motivated mainly by the desire to assist their community. The community selects the leaders because of their proven interest in serving in such a capacity.

LIVESTOCK FOR A SMALL EARTH

The Importance of Good Teaching

The trainer, or teacher, is the key to a good education program. Effective education requires that the teacher relate to people's potential and to their capacities for self-development. The teacher should avoid reinforcing feelings of inferiority or low self-esteem. Learner and teacher need to search together for alternatives and answers, be open to discovery, and have a healthy respect for local wisdom and values. If this is the attitude of the helper, and the goal is not to "teach one person to fish" amongst thousands, but to work for better conditions for the whole fishing community, then teaching is a significant part of the transformation process.

Six

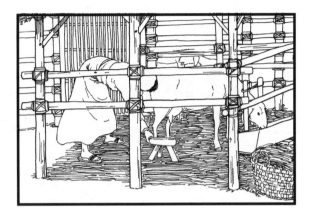

Helping People Through Livestock Projects

Many well-meaning people have designed elaborate systems for impoverished peasants only to find their well-laid plans come to naught. The belief was that traditional farmers were stubborn and could not recognize a good thing when it was offered to them. But a different picture emerged as development professionals studied how these farmers make decisions. Traditional farmers know that they should always take the pig out of its "poke" and examine it before buying it. The phrase "don't buy a pig in a poke" came from the scam of actually selling two or three cats in a bag instead of two small pigs.

Traditional farmers may be skeptical of high-yielding varieties of corn and beans that require inputs (chemicals, irrigation, etc.) that they cannot afford. If farmers make compost less efficiently than the textbook method expects, it is usually because labor demands are too great at the recommended time to follow the instructions exactly.

If farmers are given choices and weigh the available resources, they usually make wise decisions based on their own survival needs. Those involved in agricultural development should not make decisions for farmers. Rather, the role of an outside development agency is to expand the range of choices and knowledge available to the farmer for making decisions. Farmers know the local situation better than an outside agency.

Outside agents should work with the farmers in all aspects of projects, from the formation of their organization to evaluation of the program. They should also involve all who will do the actual physical work in every aspect of the program. As mentioned earlier, this often means paying particular attention to women, because of their important role in the care of livestock. An example of this can be seen in a feasibility study done on the use of Zebu cows as draft animals in Somalia. The men stated unanimously, "It is against our tradition for our women to use milking cows for cultivation, and they will therefore refuse!" But the women have to do all the tillage work while the men herd camels, and the women agreed to try anything that would help establish their crops in a timely manner. The milking cows were so tame that they could be harnessed to work with almost no training. When they were told what the men had said, the women quoted a proverb that translates roughly as, "Men talk a lot, but we women get our way."

Once people have determined what needs they want to address, it is important to select the right activity, project, or technology. Again, this needs to be done with full participation. This might involve education and dialogue with an outsider who may have information about the technology or livestock the people are considering. What follows are a few crucial principles for successful livestock projects.

All Things Are Considered

In choosing appropriate technology, including the type of animal to be raised, the farm's ecosystem must be considered. We have already referred to holistic thinking and the interrelationship of all things. A change in one part causes changes in other parts. Some changes are positive and others negative. Unfortunately, people commonly focus only on hoped-for benefits, and they may fail to foresee the possibility of unintended, sometimes negative, results. They might experience undesirable consequences if they do not examine all the facts at their disposal in the beginning.

Discussions about a planned pig project, for example, should include consideration of the high demand for grain and crop byproducts that will be involved. The farmer will have to invest resources (money, time, materials) in the enterprise. The participants need to ask, What impact will this have on the farm? What opportunities will it present? How will

the family market the pigs? How will they distribute their work? Who will make those decisions? Who will benefit? Will grain be fed to pigs that could be used as human food? Will more maize production cause increased soil erosion and lower soil fertility? Will the available manure be used to build soil fertility?

The Choice of Technology

One important criteria in choosing technology is whether it will make the total system more sustainable 5 to 10 years from now. The "system" in this case refers to the farm, but "system" should also be thought of in terms of all the natural resources available to the community.

One successful project aimed at improving the environment

through livestock and agriculture is a system developed at the Mindanao Baptist Rural Life Center in Kiniskusan, Davao Del Sur, Philippines. Called the Simple Agro-Livestock Technology package (SALT), this system uses goat-based agroforestry. Forty percent of the land is used for agriculture, 20 percent for forestry, and 40 percent for livestock/fodder. Families are producing food and income. Animals eat the fodder, and animal and green manures fertilize and build up the soil, minimizing erosion. In addition, contours, fodder banks, and tree hedgerows aid soil and water conservation.

It is preferable to choose an animal that can eat what is produced on the farm. Trees might be considered as a desirable planting to help feed ruminants, provide mulch, help control erosion, and supply fuel for cooking. Nitrogen-fixing trees will do all this, plus provide nitrogen for soil and plants.

Consideration of environmental impact should lead to a search for alternatives that mimic what nature has always been capable of providing. While some would argue that limiting the use of chemical inputs limits productivity, experience with farming methods that involve low external input are showing these to be the most feasible ones for small farmers. In most low-income countries, oxen are a much more appropriate technology than tractors.

In Guatemala and Honduras, many *campesino* families have readily adopted cross-bred dairy goats, using a zero-grazing system for the first time. They benefit from significant amounts of organic fertilizer by composting excess farm byproducts, manure, and urine. Recently their governments' inability to subsidize chemical fertilizer has made the use of organic fertilizer even more attractive to these families. As the organic content of soils grows, water retention increases and soil erosion decreases.

Many peasant families in Africa are planting forage-producing, nitrogen-fixing trees. This provides both more and better-quality forages during the usual six-to-eight-month dry season. In addition, the branches are used for firewood, reducing the need to cut other trees or purchase charcoal.

Starting with the Basics

In planning small-scale livestock projects, Heifer Project International starts by identifying the most fundamental constraint to productivity. This is what Savory (1988) calls "the weak link." Only when these "weak links" are addressed can the people and their environment be developed to their potential.

In many cases, the fundamental constraint turns out to be something the women are concerned about. For example, production from dairy animals depends heavily on ample clean water. But obtaining water is the responsibility of women, whose workload is usually such that they can only collect more water by not performing some other jobs. Either the women's labor or the animals' water needs will be compromised. By identifying the water constraint early on, alternatives may be considered to provide some additional assistance to the women. This could be in the form of draft animals to transport water. Donkeys, for example, are hardy, adapted to limited water and feed resources, and can be handled by women and children.

It is very common for people to want "the best." In a farmer's mind, this may mean new or sophisticated technology or an animal with the highest genetic potential. But "best" should be defined in terms of the local situation. A Saanen goat capable of producing 10 liters of milk per day in Britain may not be the best animal for Liberia, especially if it suffers stress in the less than optimal conditions of the tropics. A native West African goat may be best in that situation. Cross breeding is usually preferable to the introduction of purebred exotic breeds.

Starting with the basics means improving animal nutrition and health, for without these, genetic improvements will have little impact. It may, however, be desirable to introduce a different type of animal. For example, changing to zero-grazing of ruminants, which is labor intensive, is only justified when the farmer has an animal that will respond significantly to better nutrition.

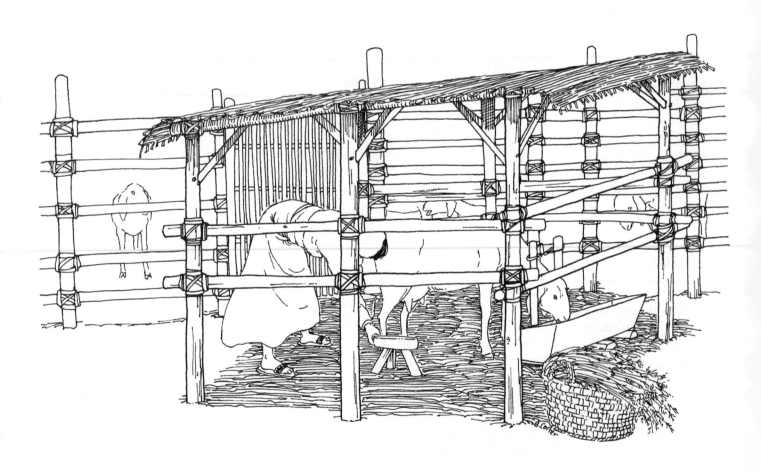

Think Small and Long-Term

It is important to start small and work toward long-term success. Many efforts fail because too much is done too fast. If possible, it is desirable to experiment with a new idea or technology before risking significant investment. This is the principle of "learning what works before expanding." In some cases, local progressive farmers may have already demonstrated proven ways of making improvements.

In Kenya, some progressive farmers pioneered the technology of zero-grazing dairy cattle. Once they had shown that milk could be produced successfully in this hot and humid area from crossbred animals, it was possible to fund small dairy projects with limited-resource farmers in this area and in other parts of East Africa.

Starting small is a hedge against the risks and the unpredictable nature of farming. Starting a project itself is a unique event. The outcome is not certain, and starting small allows for the learning and flexibility necessary for later changes. Usually it is recommended that the initial assistance to a project be limited to perhaps 10 to 30 families per year.

A development agency with a long-range view will think of success in terms of positive results that can be obtained 5, 10, or even 15 years after initiating a process. Learning from experience has much to do with effectiveness in human development programs.

The philosophy of many smaller nongovernmental organizations on the matters of size and scope differs somewhat from that of some of the larger funding organizations. Large donor agencies may be interested in fast results and may try to cover a bigger target population.

Significant change requires significant effort. But learning from small projects can later lead to slow but sure expansion. Once success is achieved, it is relatively easy to expand and replicate; but initial proven effectiveness is crucial.

Rabbit projects have relatively high rates of failure, due, in part, to an underestimation of how disciplined farmers have to be in applying good nutrition and hygiene to the raising of rabbits. In Cameroon, rabbit production was slowly introduced to small groups of farmers over several years. They needed to learn to identify medicinal plants, design cages made of locally available materials, and properly process rabbit meat. Once this was accomplished a solid foundation supported future expansion. Since then, over 1,000 farmers have begun rabbit production successfully, many without any outside assistance.

Mutual Help

Mutual aid is very common in rural areas around the world. Among the Ayamara of Peru and Bolivia, shared labor is referred to as *ayne*, while the Quechua of the Andes call it the *minka*. In Haiti it is *kombit*, in Tanzania *ujama*, and in the Philippines, *bayamhan*. Working together builds solidarity, which is one of the foundations of human transformation.

Certainly more long-term success occurs when individual families raise animals on their own farms. This assures that each family knows its responsibility and sees both immediate results and long-term rewards for their efforts. Nevertheless, we do not undervalue the importance of community solidarity. Anyone interested in development puts a high value on working together for the good of the whole community.

LIVESTOCK FOR A SMALL EARTH

Goal Setting

Planning is crucial to the success of any project. Without well-thought-out goals, groups have no basis for later evaluation of their projects. And planners should understand the terms being used in their discussions. As stated in HPI's *Planning Guide for Small Scale Livestock Projects*, "The literature on planning and evaluation does not provide any absolute definitions of terms. However, for effective communication it is important that people who work together in projects all have approximately the same idea as to the meaning of the terms they use."

The following specific planning definitions may be helpful:

- Needs—A lack of something desired by the community. They should be defined by the people involved in a given situation.
- Purpose—A broad statement of why the project or organization exists. The organization or project exists to fulfill the defined needs.
- Policy—Gives the basis for "the way we operate."
- Assumptions—Conditions, both external and internal, assumed to exist that will affect the project either positively or negatively.
- Goals—"End results" statements of what will be accomplished.
- Objectives—Specific accomplishments that are needed to fulfill the goals.
- Action Plans—Activities that will be carried out to fulfill the objectives. These are sometimes called strategies.

In discussing goals, planners should keep in mind the following considerations:

- Goals should be clear, measurable or observable, and attainable. Participants should be able to evaluate them.
- Goals should include what is to be done and what change is expected to result.
- Goals should include an estimated time period for accomplishment.
- Goals should be stimulating and motivating.

Various aspects of a sound livestock project should be assessed during the planning process. Many of these concepts, such as participation, gender concerns, sharing resources, and integrated animal agriculture, are covered in this book. Appendix 2 presents these "cornerstones" in succinct form.

The Action Plan and Calendar of Activities

An important part of planning the project consists of writing ideas into a realistic action plan. The proposed actions in the plan have to be realistic. Such a plan should define all key activities. These could include obtaining seedlings; planting pasture; training; constructing animal pens; purchasing, preparing and transporting the animals. The plan should also state the resources needed for the project, including funds, materials, and human resources, including labor.

Another important component is the timeline. When will each activity be initiated and completed, and what are some important checkpoints along the way? How and when will group leaders be elected, when will they meet, and how will decisions be made and implemented?

Very specific work plans can be written, including the dates by which certain activities will be started and finished (e.g., when construction of pens will start and finish). Experience shows that most groups have a tendency to make plans that are overly optimistic in terms of what will be achieved by a specified date. A good plan will contain opportunities for regular review of progress to allow the persons responsible to decide on appropriate corrective action. Assignment of specific responsibility assures better performance by group members and greater accountability by all involved.

The group must understand that there are certain prerequisites within a plan. Some activities have to be achieved before the next step is taken. It is important that these prerequisites be agreed on and clearly noted. For example, a plan might include such prerequisites as, "Farmers are to complete training on housing before beginning the construction of the animal pen," and "Animals will be delivered only after adequate fodder is established on the owner's field."

Continuing Contact

Regular contact with the farmers is important throughout the duration of the project. Inevitably problems will arise or new opportunities will emerge. Farming is dynamic and creative. A disease may break out, or a new resource may become available. Such situations present a "teachable moment"—a time of peak interest when relevant and timely information can be used effectively.

Independent verification can be done by the extension worker, a supervisor, or a village leader. The achievement of established milestones should be checked. The group is usually eager to obtain all the assistance promised and may be too optimistic about achieving certain deadlines.

An organization with widespread membership often feels pressure to try to help all its members or spread the benefits to all areas where it operates. This syndrome of "wide dispersal" is particularly true for some church-related organizations. Under these conditions participants may have difficulty meeting together, sharing labor, exchanging breeding males, and learning from each other. Can clustering and working with a small number of families in an accessible geographic area facilitate regular contact and reduce expenses? To benefit from group activity, participants might be best served by forming small groups that can work easily together.

Record Keeping

The most basic and necessary records are those kept by the farmer. Information on the acquisition of animals, breeding, reproduction, and current inventory should be recorded and kept up to date. Keeping this information in a simple notebook is adequate, but if this information is not recorded at the time of occurrence, it is difficult to remember dates and numbers later. Many farmers do keep considerable information in their heads, or noted on a calendar, though this is usually not enough for use in management of their livestock.

Many projects attempt to put information on record cards and notebooks, and this is sometimes successful. A study of farmers in one small-holder dairy project in Tanzania found that only 45 percent of the farmers were keeping written records—and that is probably better than what occurs in many projects. The country director for HPI in Tanzania remarked, "There are card systems working extremely well in private farms (e.g., in Canada), but I have not seen a card system successfully implemented in development projects." One exception might be the Indo-Swiss Project in Andhra Pradesh in India, where the cards were kept up to date for the most part.

Record keeping should include the following steps:

- Compilation of information
- Analysis
- Conclusions
- Decisions
- Implementation of decisions

Some important areas of record keeping include the following:

- Herd or flock inventory—Overall growth or decrease in herd numbers as a result of deaths, births, sales, and slaughter (or loss due to other causes). These numbers are recorded for the animals in each category, (e.g., bulls, cows, heifers, calves). Ideally, the farmer should also record weights at birth, weaning, and slaughter.
- Breeding records—Information on each breeding animal, including heat cycles, date bred, sire, number and date born, sexes of offspring, and number weaned.
- Income and expense—Costs of starting up a project (investment), as well as operating expenses, (e.g., feed, health care, supplies, etc.). Obviously, the record of income from the sale of milk, meat, wool, or other products is important.
- Health records—Vaccinations, parasite control, and disease treatments.
- Land and Soils—Quality and quantity of hay, soil test results, fertility, and fertilization.

When simple records are kept concerning normal costs such as feeds, medicines, and labor, it will be fairly easy to determine if the project is making or losing money.

Most practical livestock production manuals, some of which are listed in the Bibliography, include ideas on record keeping. These references usually provide sample forms that livestock project personnel can use or adapt. Heifer Project International has developed calendars and herd management wheels for use in a number of projects. These help keep track of what activity should be done on a particular day.

Project managers and extensionists also need tools for keeping track of various activities during the execution of projects. Ledgers can be devised for any aspect of a project for which an ongoing record will be useful. These could include such activities as training, passing on the gift, meetings, and so forth.

In addition to the above, it might be helpful to record some basic information on each family before they are provided with an animal. This baseline information will greatly assist in later determination of the impact of the project on its members.

Heifer Project International has found it very helpful to groups if they record some of their plans and progress graphically on posters placed on the walls of their meeting location. For example, if a group posts its goals and activity calendar, they can easily review them each time they meet.

Monitoring and Accountability

Keeping adequate information and reviewing progress on a regular basis are extremely important. Reports provide written information for learning and decision making. They also allow leaders to be responsible to members and the project organization to be accountable to those assisting them. Reporting must be included in the initial project plans. Most project organizations carry out regularly scheduled meetings to review progress. Often quarterly and annual meetings are useful for more thorough evaluation and replanning for the next period. Obviously, clear objectives and action plans will greatly facilitate such review. Appendix 3 provides a listing of various forms and guidelines that livestock projects can use for monitoring and reporting.

Seven

Passing on the Gift

The Chain of Being, for instance—which gave humans a place between animals and angels in the order of Creation—is an old idea that has not been replaced by any adequate new one.

—WENDELL BERRY

Sharing and caring are profound human responses. People who share are interested in the needs of others. These responses are crucial to the process of sustainable development for two reasons. First, sharing with others protects and enhances dignity and self-esteem, especially when the one who was a recipient becomes a giver. Secondly, sharing can be contagious, and it perpetuates material, social, and spiritual benefits—it creates a chain reaction. Developing the capacity and motivation to work for the good of others in the community should be a goal of any development project. Sharing promotes the spirit of community building and is at the core of human transformation.

People in need who are helped by others should be given the opportunity to partake in that spirit of sharing. We call this principle "passing on the gift," and it has been the trademark of HPI since the organization's founding. It is a primary reason for the relatively high rate of success in multiplying benefits to project participants.

Experience has shown that helping others is transformational for both parties. It is for that reason that HPI continues to foster the idea of livestock as a "gift." Receiver and giver are blessed and changed. A widow in Uganda stated that she could not understand how a person in the United States could help provide her with a cow when her experience was that her own government had denied similar help to her community. Having received such a gift, she was now eager to "bless others as she had been blessed." A project of this type can transform a whole community, as hope, mutual assistance, and a spirit of caring for each other are enhanced.

A Maasai woman in Kenya whose family received cattle after theirs were lost in a drought stated, "You have added to our strength." The Maasai, who consider cattle an important part of their way of life, readily understood the importance of spreading the benefit to others.

LIVESTOCK FOR A SMALL EARTH

Every Participant a Provider

Heifer Project International pioneered the concept of passing on the gift. Simply stated, passing on the gift means that those who receive a female animal—the "living gift"—are required to pass on one or more offspring of equal value and kind to another family. Those who have resources, in this case farm animals, should share this food-producing resource with those in need. Those in need might be neighbors next door or across the world. Passing on the gift was one of the pillars upon which the original idea for "Heifers for Relief" was built back in the 1940s.

Heifer Project International continues to share animals and training and expects those who receive to become providers of livestock and training to other people. It has been interesting to observe that in the dozens of countries where this principle has been applied, people of many different cultures and histories readily accept it.

Participants extend this "chain of life" and pass on animals as living loans in numerous ways. The gift might consist of several young pigs, or five offspring from one doe rabbit, or the first female calf from a water buffalo.

The Akha people of northern Thailand actually pass on the second calf and return the original buffalo cow after the third calf is weaned from its mother. It is an arrangement they decided on themselves, even though it actually exceeds the minimum requirement of the original idea—that a person share something of at least equal value as that which was received.

In the Dominican Republic, local project committees require the return of a second offspring, either a bull or heifer calf, to support the ongoing needs of the farmer associations. This is another example of sustainability—in this particular case, the ongoing sustainability of the organizations.

In Tanzania, project organizations decided to require the return of two offspring. The first female calf is for passing on to another family, while another calf, which could be a male, is returned to the project to help cover ongoing training and follow-up costs. Thus, project sustainability is guaranteed and locally paid for.

Passing on the gift does not only apply to sharing animals. It also means sharing knowledge and skills gained through training. A development project provides opportunities or funding for training so that practical skills and knowledge can be learned in settings close to the real-life situations of the participants. People who receive training pass on the gift of new knowledge and skills to other farmers in their own or other communities.

Development work should help people search for an approach that facilitates expansion of benefits to a larger number of people. People must not be treated as objects, as a "target group," or the "beneficiaries" of assistance. All involved are subjects and responsible partners. The concept of passing on the gift teaches all of us to be good stewards of our own gifts and of the resources entrusted to us.

A Carabao Bank for Small-Scale Farmers

Xavier University in Mindanao in the Philippines has been assisting area farm families since 1959. The "carabao project" began in 1990, when HPI provided seed funds to the university extension service through a local credit cooperative for loans to project farmers. Five-year loans at a 12 percent annual interest rate enabled each farmer to purchase one carabao (water buffalo).

In this area over 90 percent of the population is rural and depends on crops and animals they raise for their food. The small-scale farmers consider the carabao as their most useful and important economic resource. Aside from providing draft power for producing rice and other crops, carabaos pull carts, carry water, produce manure for the garden and field, and represent a form of savings. One of the primary advantages of the carabao is that it thrives on a diet that consists mainly of crop by-products and local forages.

Lacking a carabao, farmers must use costly or inefficient methods to cultivate their plots. Using hand tools and employing all available family members is about 15 percent as efficient as using a carabao, and this approach cannot bring enough land under cultivation for a family's food needs. Families often must borrow money at high interest rates to pay for tillage. Using the local *alima* system, a farmer without a carabao would have to pay one-third of his total crop production in return for a loaned carabao to the owner of the carabao.

The project's training programs begin before distribution of the carabao and include such topics as community organizing, care for the soil and environment, feeds and feeding, animal health care, and breeding. Since the project's inception, 15 self-governing farmers' groups have been formed, composed of five farmers each. They manage the collection of loan payments and all of the training activities.

Forty limited-resource families currently participate in the project. Through repayments the farmers plan to purchase 35 more carabao over five years. Timely repayment by farmers will ensure that their neighbors receive an animal through the pass-on program. After five years 75 families, or 450 people, should benefit directly from this program through increased food production and a better quality of life.

Examples from Korea and India

One of the most notable examples of passing on the gift comes from the Republic of Korea. Farmers who received dairy cattle from HPI in the 1950s and 1960s have continued passing on one heifer calf for each one received for all the intervening years. During these years the standard of living for most Koreans has improved greatly due to the country's economic growth, and these farmers feel they are able to pass on the gift in a more significant way. These successful farmers and businessmen recently began working with a new Korean organization in partnership with HPI/USA. The aim is to enable farm families around the world to improve their standard of living. In 1989, 15 of the Korean farmers gave $12,000 to help dairy farmers in China. In 1991, they gave an additional $3,000 for tribal families in northern Thailand.

In India funding agencies for livestock projects work with the banking system to back up loans to farmers. Typically the agencies provide collateral equal to 15 percent of the total loan to a local bank, where it earns interest at competitive rates. The farmers sign for a loan against the collateral. They put their name or thumbprint on the contract and start the process of establishing their credit rating—usually for the first time in their lives. If the people are tribal farmers or landless, the state government often provides a 25 to 50 percent subsidy directly to the bank, reducing that farmer's loan amount. To qualify for the loan, farmers must also purchase livestock insurance worth 5 percent of the value of their animal. Loans are provided at an annual interest rate of 4 percent—lower than the usual 12 percent rate. These interest rates are also much lower than the rates charged by local (unofficial) moneylenders, which can exceed 100 percent a month.

With the possible subsidy of 50 percent, the low 4 percent interest rate, and, in some locations, a one-year grace period before repayment must begin, many farmers also agree to pass on a female offspring to their neighbors. The animals are passed on within four or five years, after most of the loans are repaid. The bank then releases the collateral plus the interest it has earned. This money can be used again to fund another project.

Most of the loans are repaid within the agreed-upon time—which initially surprised the bankers, who expected that limited-resource, landless farmers would be very high credit risks. The excellent repayment rates have made the bankers cooperative and enthusiastic. One bank manager said the farmers had a better repayment rate "than the owner of the local soybean oil processing mill!"

Heifer Project International began to provide collateral funds for loans in India in 1983 and tracked these loans very carefully. By 1988, the program began to receive funds from released collateral, and this money was used to help many more families than the budget would otherwise have allowed. Over the three-year period from 1988 through 1990, this system leveraged $298,000 from the $168,000 actually spent on livestock projects during those years.

Passing On: How to Deal with the Different Species

Another form of passing on the gift is to provide funds from the sale of an animal or animal products. In some poultry projects, many families use income from egg sales to purchase an equal number of day-old chicks for another group.

Some fish farmers sell part of their harvest for money to buy fingerlings for their neighbors. If the species of fish is one that can be reproduced in ponds by the local farmers and no outside source of fingerlings is required, then the actual fingerlings are passed on. Since fingerlings are usually abundant and do not actually represent a significant pass-on in themselves, the farmers in aquaculture projects usually also provide labor to help their neighbors dig ponds. These ponds are used for raising fish and for harvesting water for home and garden use, especially during the dry season.

Farm families who receive honeybees and hives usually pass on hives to their neighbors. This requires that they sell honey or other products such as wax or royal jelly, and purchase some special boxes for a hive to be passed on. When the original bees reproduce, an additional queen bee is provided with the hive to the recipient farmer.

Heifer Project International has developed general guidelines for passing on animals. Large livestock, such as cattle, water buffalo, camels, yaks, alpacas, llamas, and donkeys, are usually passed on as one mature offspring for each animal received. Smaller, litter-bearing animals, such as swine, sheep, goats, rabbits, and guinea pigs, often require that more than one offspring be passed on for each original animal received by a family. There may be exceptions to these general guidelines, however, depending on the local situation. There may also be differences within the same species. For instance, some breeds of swine give birth often and have very large litters, perhaps up to 12 pigs at a time. Others give birth less often and with fewer pigs per litter.

Local project committees are usually responsible for helping people who have received livestock share their animals' offspring with others. These committees help select new participants and make certain the new recipients are prepared to care for the animals and continue the "chain of life." Requests for support sent to HPI must explain, in writing, the agreement of the local group and individual participants to pass on the gift. This contract, or agreement, is signed by the group and each participating family.

Constraints and Challenges

People may think that passing on the gift means that 100 farmers who have received 100 water buffalo heifers will pass on 100 buffalo heifer calves, maybe within a year or so. Others may know that when the calves are born, about half will be heifers and half will be bulls. So only about 50 heifer calves might be passed on in the first round. A water buffalo may not have its first calf until it is four years old. Depending upon several management factors, they may have a period (known as the calving interval) of two or more years between the births of calves. For most dairy cattle, the age at first calving is around two to three years, with a calving interval of 12 to 24 months.

These factors all reduce the normal rate of passing on the gift from that ideal goal of 100 percent. Clear understanding between HPI and its partners, good relationships within the community, and adequate technical support are all important to successfully passing on the gift.

Both technical and social problems may arise in the course of a program. Some technical causes of slow or low pass-on rates include (1) poor farm management, such as dirty housing or lack of ventilation; (2) poor animal nutrition; (3) poor animal health care or a disease problem beyond the control of the farmer; (4) poor breeding rates because of feeding, management, and/or disease. Social problems may include (1) poor communication or understanding; (2) power struggles, breakdown of unity, and/or jealousies; (3) poor organization or management; (4) negative outside forces, such as peace and order problems; (5) animal losses due to poisoning, theft, or other causes.

Problems usually become more severe and occur more frequently during second and later pass-ons. This is mainly because the recipients of the later pass-ons may not have been included in the early stages of decision making in a project. A participatory process in which the community members all feel ownership and have had a hand in designing the program can avoid these problems.

Problems and Solutions in India

Difficulties encountered with livestock projects in India provide valuable insights. One problem occurred when participants did not adequately care for young animals destined to be passed on. These farmers were poor and hardly able to care for their own children, let alone an animal that was to be given away. Heifer Project International now approves projects later in the process of a community's development, after extensive community work and detailed explanations to recipients. This has improved the pass-on rates. Project managers now also keep a list of farm families who have been selected to receive the pass-on animals. This provides an additional social factor: the original farmers know who will receive the offspring of their animal, and the "second-generation" farmers know where their animals will come from.

A second problem is that many of the participating families are so poor that any unforeseen change in their meager income or health may require that they sell or slaughter the animal. These families have very limited resources, are usually landless, and often have no form of security other than their animal.

In India, where it is possible for HPI to provide development assistance to a group of such poverty-stricken people through bank loans, the bank actually owns the animals until they are paid for by the farmers. Of course, project committees work hard to make sure that farmers are well-motivated and show good indication of ability to repay the loan. Insurance is purchased for the animals in case of accidental death, and loan payments are made through the local cooperative marketing association over a period of four to five years. When the loan is repaid, the farmer has established his or her credit rating with the bank and does not require the outside agency to provide collateral for a future loan. The released collateral is then passed on for another community or group of farmers to obtain loans for the purchase of animals.

The Challenge: That More Might Benefit from the Gift

Each one, as a steward of God's different gifts, must use for the good of others the special gifts they have received from God.

—I PETER 4:10

One of the greatest challenges in the development field today is to use very limited resources in a world of such great need and poverty. The passing-on-the-gift approach offers one answer and has proven effective when applied to livestock development projects in many circumstances.

Can this spirit of sharing and this system of expanding benefits to more and more people be applied to others areas of development as well? Adaptations of this principle certainly can be used in many types of projects that work with people on basic human needs. Housing, forestation, and seed banks are some examples.

The significance of sharing and caring through the pass-on system has been manifested in countless ways and places. Many of the participating projects and communities celebrate special days for the ceremony of passing on the gift. People gather in the village plaza in Latin America, or perhaps under a large acacia tree in the African savanna, to present animals from the original owners to the new families. The event often involves speeches by community leaders and invited guests. It is an opportunity for the community to invite area livestock officers, veterinarians, bank managers, and government officials to participate in their activities. This brings about a better understanding between people and can enhance relationships all around. It is a time of pride and joy for the farm families who raised the animals, and it is a time of great anticipation for the receiving families.

A successful passing-on-the-gift program shows an accepting attitude among the participating families. It often indicates true feelings of participation and ownership of the whole program. Passing on the gift can potentially benefit the entire community, not just a few.

CHAPTER
Eight

Behind These Animals Is a World of Hope

In this book we have stressed the importance of animals as a part of sustainable agriculture for small-scale farmers. More importantly, we have emphasized that people are the subjects of their own development. Sustainable development will happen only when people participate in shaping their own vision of the future.

Rural families, whose tradition has been to live in harmony with the earth and its creatures, are threatened on all sides. Farmers all over the world are facing the extinction of farming as they and their ancestors have known it. Agriculture that respects life, builds community, and provides for the physical, social, and spiritual needs of people is being transformed into a nonurban industry.

The contrasts between large-scale livestock confinement systems and holistic approaches to farming are stark. In the United States, agriculture is moving from small-scale, broad-based family farming to large-scale, industrial-based farming. Production of cattle, chickens, and, more recently, hogs, is increasingly carried out in huge, vertically integrated enterprises. The farm family is only a small cog in the system, if, indeed, they play any part at all. Rural areas are becoming, as Wes Jackson puts it, "a dispersed minority with no political clout—marginal isolates rapidly becoming like Third World people" (1987).

W. C. Lowdermilk composed the "eleventh commandment" and wrote in his classic work, *Conquest of the Land Through Seven Thousand Years*, the following reflection:

The Eleventh Commandment

Thou shalt inherit the Holy Earth as a faithful steward, conserving its resources and productivity from generation to generation. Thou shalt safeguard thy fields from soil erosion, thy living waters from drying up, thy forests from desolation, and protect thy hills from overgrazing by the herds, that thy descendants may have abundance forever. If any shall fail in this stewardship of the land, thy fruitful fields shall become sterile stone ground and wasting gullies, and thy descendants shall decrease and live in poverty or perish from off the face of the earth.

I wonder if Moses had foreseen what was to become of the Promised Land after 3,000 years and what was to become of hundreds of millions of acres of once good land such as I have seen in China, Korea, North Africa, and the Near East, and in our own fair land of America. If Moses had foreseen what suicidal agriculture would do to the land of the holy earth, might he not have been inspired to deliver another Commandment to establish [people's] relations to the earth and to complete our relations to each other, and to the holy earth? (Freudenberger 1990, 72).

The situation for small-scale farmers in low-income areas of the world is even more desperate than it is in high-income countries. Throughout the world, hundreds of millions of people depend on subsistence agriculture for their food and livelihood. Many of them face diminishing production on eroded soils. They have few resources other than their family's labor force and little access to affordable credit or stable markets. Topsoil losses to farms in Iowa, one of the richest agricultural areas in the world, are more than matched by their counterparts in Peru, Nicaragua, and Nepal, where small-scale farmers struggle to scratch out a living. Many finally give up and join the mass migration to the slums of the major cities of their countries. The same scenario is repeated over and over in Asia, Latin America, and Africa.

Asking Questions and Visioning the Future

We believe the principles and methods described in this book provide an undergirding for sustainable development. Stewardship and wise management of the natural resources available to small-scale farmers are at the heart of this approach. "Sustainability" has been elevated as a priority issue in many forums around the world, pointing to a hopeful, though tenuous, future.

Power can be derived from simply asking questions. It is only when a vision is generated from below, in democratic forums and people's organizations, that there is a possibility that the vision will become reality. As Dean Freudenberger (1990, 22) so aptly put it, "Hope is born when we ask the questions about things that trouble us most. Asking questions sets the stage for new futures that, in light of the present state of affairs in our agriculture, are so desperately needed."

Finding the answers to many of the questions we must ask will be a challenge. How are we to envision a sustainable future? What should the landscape look like in 25 or 50 years? What do concerned people everywhere need to do next? These are questions worthy of much reflection and discussion.

The visions and dreams of an equitable, just, and sustainable world should not come only out of the minds and mouths of politicians, professors, authors, and program directors. The "vision question" should be asked in thousands of meetings, interviews, and conferences around the world.

The Land Stewardship Project, an organization that promotes sustainable agriculture with family farmers in the United States, asked its constituents about their vision of the future. Although the context and type of farming are different than in the developing world, the vision espoused by these farmers is instructive for people interested in issues of a sustainable future throughout the world. Here is a summary of the responses:

> The theme of dispersal ran through the commentary at the meetings; dispersal of animals from feedlots to individual farms and dispersal of people from concentrated urban areas to rural areas. The desire for greater diversity was universally raised, including a greater variety of cultivated crops and more non-cultivated species of plants and animals. Diversity also took the form of more complex rural communities in terms of race, ethnic heritage, gender, age and work. Participants dreamed of seeing improved relationships within families, among farmers, between farmers and rural community members, and between people in rural areas and the smaller urban areas. In this scenario, it would be possible for youth to return to productive and enjoyable livelihoods in their rural home communities.

We believe that nongovernmental organizations (NGOs) working in partnership at the grassroots with community-based organizations are a tremendous resource to bring about this future. For NGOs working in development, such as Heifer Project International, this means tapping into the core of values that inspired its creation a half century ago. People's organizations and NGOs are part of the answer, but only part of it; NGOs must join with official governmental and multilateral agencies to forge policies and programs that promote sustainable agriculture and a just and equitable society.

Nongovernmental organizations that want to pursue sustainable development must do more than promote their particular "niche," whether that be livestock, health, appropriate technology, or forestry. For HPI, at the minimum, the cows, goats, pigs, and llamas are means to assist farmers to attain the basic elements for survival. But we must take a larger perspective and search for ways to promote the renewal of life, the preservation of soil, and the building of communities. We cannot expect to promote animal raising without promoting the whole ecosystem or without reexamining our assumptions about the underlying causes of poverty and hunger.

Working with livestock and plants and building community in a sustainable way is a complex process. Like a spider's web, the help an NGO provides small farmers is intricately linked to the many strands that make up the whole. This calls us to a humble and continuous critique of our modes of operation and requires us to look for ways to transform ourselves as well as our own organizations. For those who

LIVESTOCK FOR A SMALL EARTH

would dedicate themselves to the mission of improving life on this planet through sustainable agriculture, we need a measure of both realism and faith for the undertaking.

Finally, NGOs, such as Heifer Project International, have a powerful resource in their volunteers and other supporters. Educating and organizing donors and volunteers of NGOs to engage in grassroots advocacy for a sustainable and just future is one of our greatest challenges. While volunteers and full-time staff will always be relatively few in number, a broadly based and informed constituency is needed if real transformation is to take place.

One of the slogans of the Sarvodaya movement in Sri Lanka is, "We build the road, and the road builds us." Indeed, it is our hope that the ideas, stories, and information contained in these pages will be of help to many as they journey down that road.

Appendix 1

About Heifer Project International

Heifer Project International is one of many nongovernmental organizations (NGOs) that work to alleviate poverty and hunger. However, HPI is perhaps the only NGO that specializes in animal agriculture as the vehicle for the development of people and communities, helping families become self-sufficient in food and income.

Heifer Project International expresses its mission as follows:

In response to God's love for all people, the mission of Heifer Project International, in partnership with others, is to alleviate hunger, poverty, and environmental degradation by:

A. Responding to requests for development assistance, including animals, training, and technical assistance which enables families to seek self-reliance in food production and income generation on a sustainable basis.

B. Enabling people to share (pass on the gift) in ways that enhance dignity and offer everyone the opportunity to make a difference in the struggle to alleviate hunger and poverty.

C. Educating people about the root causes of hunger and poverty based on Heifer Project International's experience and insight gained from working with animals in development since 1944.

D. Supporting people in sustainable development and the stewardship of the environment through responsible management of animal resources.

Hungry children were why HPI was formed in 1944. The dream was to provide families a source of food rather than short-term relief, so that families could feed their own children every day. This dream has been fulfilled with families in 110 countries around the globe. Currently, HPI projects operate in 33 countries and 15 states.

The trademark of HPI's approach is assisting communities through the "living gift" and the understanding that each person who is helped will "pass on the gift." Since its first shipment of 18 heifers to Puerto Rico in 1944, Heifer Project International has provided almost 100,000 large animals, 1,300,000 fowl, 5,000 beehives, and 1,300,000 fish fingerlings to families all over the world.

Heifer Project International works at the grassroots by providing animals and training to organized local groups that request assistance. The organization's work is community driven by the expressed needs of the people participating in the project. Typically, a project consists of three essential components: (1) livestock and other material goods; (2) training and extension work; and (3) organizational development, which includes planning, management, record keeping, passing on the gift, and reporting and evaluation.

Heifer Project International has acquired vast experience in its work with livestock projects since 1944 and has developed an understanding of the factors that contribute to effective projects. Twelve of these factors have been identified as "cornerstones" of HPI's program, and these are used to screen, plan, manage, monitor, and evaluate its programs. The 12 cornerstones are described in Appendix 2.

Appendix 2

HPI Cornerstones for Sustainable Development

Cornerstone: A basic element; foundation

Staff members of HPI, both in the United States and overseas, have developed a working set of principles that we are calling the HPI Cornerstones for Sustainable Development. These cornerstones are aspects of our program that we consider essential for effective, sustainable development of people, with emphasis on families. Also of primary importance to HPI is the well-being and productivity of the livestock, and regeneration of the environment.

The HPI cornerstones form the basis of our Accountability Process. All organizations and farmer groups are screened, monitored, and evaluated according to these cornerstones, and the project plan made by the groups takes these cornerstones into consideration.

The 12 cornerstones fall neatly into the acronym **Passing on the Gift**—which is itself a cornerstone—making the list easier to remember for presenters, volunteers, trainers, donors, board members, and partners.

P articipation/Cooperation
A ccountability
S haring/Caring
S elf-reliance
I ntegrated Animal Agriculture
N utrition and Income
G enuine Need

on the

G ender Concerns
I mproving the Environment
F amily Focus
T raining and Education

Participation/Cooperation: HPI works with grassroots groups or intermediary organizations representing grassroots groups. A truly effective group will have strong leadership, committed to involving all members in decision making. All members of such groups will cooperate in the group's activities. Heifer Project International's requirement for regular reporting and group-level decision making means that an appropriate structure to manage the responsibilities must be in place. These responsibilities include overseeing that each farmer is fully prepared before animals are delivered and choosing which families in the community will receive animals.

Accountability: The group defines their own needs, sets goals, and plans an appropriate strategy to achieve their goals. Heifer Project International provides guidelines for planning the project, screening recipients, monitoring farmers' progress, and conducting self-reviews. Careful financial management and record keeping of group activities are required by HPI for each project. Progress and financial reports are submitted to HPI every six months, and each project undergoes an evaluation toward the end of its agreement with HPI so that the lessons learned can be shared with other projects.

Sharing and Caring: Caring about those in need and sharing ourselves with others is the glue that binds everyone who is involved in HPI's work. Heifer Project International sees its work as a process of transforming attitudes and relationships. The type of "development" to which we aspire positively affects the whole person—socially, spiritually, physically (economically), and mentally. Heifer Project International screens the existing groups who wish to become our partners to make sure that a concern for others is demonstrated in the organization's activities and/or bylaws.

Self-Reliance: Since HPI can only fund a project for a limited time, sponsored groups must plan to support themselves eventually. Heifer Project International has found that this self-reliance is most easily achieved when the group works on varied activities and looks for support from several sources. Groups that encourage their members to share ideas and try new approaches are most likely to survive a long time. Heifer Project International encourages groups to seek the recognition and cooperation of local community or village leaders. The practice of passing on the resources provides an important means of sustaining group momentum and activity over the long term.

Integrated Animal Agriculture: Feed, water, shelter, reproductive efficiency, and health care are the essential ingredients in successful livestock management, and they must be available for the livestock provided by HPI. Animals and plants belong together in an integrated system in which manure, urine, forages, and crop residues are recycled. Heifer Project International ensures that the livestock do not worsen any critical problem the farmers have, such as water availability, labor, and land use problems. Finally, the group activities should help to strengthen family farming and rural communities.

Nutrition and Income: Nutrition for the family is a high priority for HPI, just as it is for the families that receive HPI animals. Livestock contribute directly through protein (milk, meat, eggs) and indirectly through draft labor for crops and manure for fertilizer. Livestock can provide long-term economic security for education, health care, housing, and emergencies of all types. For this to take place, a viable market must be in place for any surplus livestock products. Both short- and long-term income potential are considered in the project plan.

Genuine Need: For HPI, the criteria for selection of families to receive animals give priority to disadvantaged people who have the minimum required resources (for example, malnourished families, the unemployed, widows, ethnic minorities, the handicapped). In making the final decisions as to who should receive animals, the group members themselves determine who are the needy in their community. The poorest in the community should be included in the group membership.

Gender Concerns: Heifer Project International encourages women and men to share in the decision making and to share the benefits of the projects in ways that are culturally appropriate. Both men and women need to be involved because the concerns of women may be different from those of the men in the group. In many families, the women are the ones taking care of the animals, and their input is important to the success of the projects.

Improving the Environment: The introduction of the HPI livestock should improve the environment by having a positive impact on one or more of the following: soil erosion, soil fertility, sanitation, forestation, biodiversity, pollution, wildlife, and watershed condition. In addition, the livestock should not cause or worsen any environmental problems.

Family Focus: Heifer Project International prefers to support projects in which the whole family participates and is strengthened by this participation. Not only should all family members see benefits from the project, but also the community should benefit from and be strengthened by the project.

Training: The group decides on their own training needs, and local people are involved in carrying out the training to the extent possible. Training includes formal sessions as well as informal (farm visits, demonstrations, model or promoter farmers) and is "hands-on" more than academic. Training for technicians and farmer leaders includes training in effective teaching methods. In addition to technical training for production, HPI groups are assisted in leadership skills and in developing an effective group.

Passing on the Gift: Passing on the gift embodies HPI's philosophy of practical sharing and caring. Passing on livestock offspring and knowledge gained through training means that HPI provides "living loans" that people will eventually pay back. Successful completion of a pass-on contract provides a sense of pride and dignity that is not possible with a simple handout. The groups of farmers themselves make the decisions about how to pass on the gift and to whom the animal is passed on, using guidelines provided by HPI. The only requirement is that the plan be realistic and fair.

Appendix 3

The Accountability Process

There are many components in HPI's accountability process. First is the selection of projects using established criteria and involving careful planning by the project holder. This is followed by semi-annual reporting, monitoring both finances and progress toward the goals and objectives set by the groups. The groups also conduct their own self-reviews (evaluations) on a regular basis. Every country/regional program regularly undergoes a thorough review by a team including the responsible program director from headquarters. Periodically, case studies of a sampling of projects collect more qualitative, intangible information about the projects.

The above process aims to ensure that all projects are administered in a way that wins the trust and confidence of both the farmers in the groups and the HPI donors. The process also allows HPI to learn from experience. Learning feeds into a comprehensive training program in project areas and provides education and feedback for HPI supporters in the United States and other countries.

The HPI accountability process is based on the 12 HPI cornerstones (see Appendix 2). At each stage in the process, consideration is given to all 12 cornerstones, although some take on more importance at different stages. For example, "Genuine Need" is important in the screening of groups (to ensure that the group works with the people who have few other resources), while "Improving the Environment" is important in monitoring the progress of each project.

The staffs' role in the accountability process involves facilitating each groups' own development rather than judging success or failure. For this reason, emphasis is given to the training of groups to manage their own monitoring and evaluation.

The process also allows HPI to be fully accountable to donors by collecting all relevant information from projects, such as the number of animals passed on or the number of women trained. All this information is fed into a database, which allows analysis of projects by individual project, country, or geographic area. Information from this database is published every year.

Required Forms and Person(s) Responsible for Completing Them

The following is a list of instruments used in the accountability process. Anyone interested in acquiring any or all of these may write to the International Program Department of Heifer Project International at 1015 S. Louisiana, Little Rock, AR, 72202.

Some forms, such as the three ledgers listed under **Progress Reports,** are designed to assist the project leaders and are not required to be sent to HPI. Others are designed to assist the country/regional representative in screening the projects and are also not required to be sent to headquarters. Certain forms, such as Progress Reports and Financial Reports, are sent first to the country representative, then on to headquarters for the program director's monitoring.

Form	Person(s) Responsible
Project Planning and Approval Stage	
☐ Information Request	Project holder
☐ Information Screen	Country/region rep.
☐ Project Plan with Pass-on Guidelines	Project holder
☐ Plan Screen	Country/region rep.
☐ On-Site Assessment	Country/region rep.
☐ Country Representative Screen	Country/region rep.
Progress Reports	
☐ Recipient and Pass-on Ledger	Project holder
☐ Training Session Ledger	Project holder
☐ Meetings Ledger	Project holder
☐ Progress Report	Project holder
☐ Representative Report	Country/region rep.
☐ Database entry forms	Database clerk
☐ Project Visit Report	Any HPI staff

Form	Person(s) Responsible

Financial Reports
- ☐ Financial Report   Project holder
- ☐ Database entry forms   Database clerk

Audits
- ☐ Audit Guidelines   HPI staff

Internal Reviews (self-evaluation)
- ☐ Self-evaluation Manual   Project holder

Case Studies
- ☐ Guidelines for Writing Case Studies   As assigned

Program Reviews
- ☐ Evaluation Terms of Reference   Evaluation team

Reference List and Bibliography

General

Aaker, Jerry. 1993. *Partners with the Poor: An Emerging Approach to Relief and Development*. New York: Friendship Press.

Ariyaratne, A. T. 1980. *Collected Works*. Nandsena Ratnapala. The Netherlands: Sarvodaya Research Institute.

Bread for the World Institute. 1990. *Hunger 1990: A Report on the State of World Hunger*. Washington, DC. (Bread for the World Institute publishes an annual report on the state of world hunger.)

Breslin, Patrick. 1987. *Development and Dignity*. Rosslyn, VA: Inter-American Foundation.

Brown, Lester, et al. 1992. *State of the World 1992: A Worldwatch Institute Report on Progress Toward a Sustainable Society*. New York: W. W. Norton and Company. (The Worldwatch Institute publishes a State of the World report each year, covering many aspects of environment, population, development, and food production.)

DeVries, James. 1992. "Development or Transformation: Reflection on a Holistic Approach to People Centered Change." Little Rock, AR: Heifer Project International.

Grigg, David. 1985. *The World Food Problem*. New York: Basil Blackwell.

Harris, Marvin. 1989. *Our Kind*. New York: Harper Collins.

Jackson, Wes. 1987. *Altars of Unhewn Stone*. San Francisco: North Point Press.

Korten, David C. 1990. *Getting to the 21st Century: Voluntary Action and the Global Agenda*. West Hartford, CT: Kumarian Press.

Korten, David C., and Rudi Klauss, eds. 1984. *People-Centered Development: Contributions Toward Theory and Planning Frameworks*. West Hartford, CT: Kumarian Press.

Lowdermilk, W. C. 1975. *Conquest of the Land through Seven Thousand Years*. Agriculture Information Bulletin, no. 99. Washington, DC: U.S. Department of Agriculture, Soil Conservation Service.

Robbins, John. 1987. *Diet for a New America*. Walpolem, NH: Stillpoint Publishing.

Schumacher, E. F. 1973. *Small Is Beautiful: Economics as if People Mattered*. New York: Harper & Row.

United Nations Development Program. 1990. *Human Development Report 1990*, New York: Oxford University Press.

Environment

Barghouti, Shawki, Carol Timmer, and Paul Siegel. 1990. "Rural Diversification: Lessons from East Asia." World Bank Technical Paper, no. 117. Washington, DC: World Bank.

Ehrlich, Paul, and Anne Ehrlich. 1981. *Extinction: The Causes and Consequences of the Disappearance of Species*. New York: Random House.

Freudenberger, C. Dean. 1990. *Global Dust Bowl; Can We Stop the Destruction of the Land Before It's Too Late?* Minneapolis, MN: Augsburg.

Grant, Lindsey. 1992. *Elephants in the Volkswagen*. New York: W. H. Freeman & Co.

World Commission on Environment and Development. 1987. *Our Common Future*. Oxford: Oxford University Press.

Sustainable Agriculture

Berry, Wendell. 1987. *Home Economics*. San Francisco: North Point Press.

Beets, Willem C. 1990. *Raising and Sustaining Productivity of Smallholder Farming Systems in the Tropics. A Handbook of Sustainable Agricultural Development*. Alkmaar, Holland: AgBé Publishing.

Coppinger, Raymond, Elisabeth Clemence, and Timothy Coppinger. 1992. "Considerations for a Sustainable Society: The Role of Livestock in Sustainable Agriculture." The Land Report.

Devendra, C. 1993. "Sustainable Animal Production and Self-Reliance for the Poor in Developing Countries." Paper presented at the International Development Conference, Washington, DC.

Durning, Alan B., and Holly B. Brough. July 1991. *Taking Stock: Animal Farming and the Environment*. Worldwatch Paper 103. Washington, DC: Worldwatch Institute.

Edwards, Clive, R. Lal, P. Madden, R. H. Miller, and G. House. 1990. *Sustainable Agricultural Systems*. Ankeny, IA: Soil and Water Conservation Society.

Gips, T., P. Allen, and D. van Dusen, eds. 1986. "What Is Sustainable Agriculture? Global Perspectives on Agroecology and Sustainable Agriculture Systems." In *Proceedings of the 6th International Scientific*

Conference of the International Federation of Organic Agriculture Movements, vol. 1, 6474. Santa Cruz: Agroecology Program, University of California.

Institute for Agricultural Biodiversity. 1992. "Why Preserve Diversity?" Decorah, IA.

Jackson, Wes. 1980. *New Roots for Agriculture*. Lincoln, NE: University of Nebraska Press.

Land Stewardship Project. Autumn 1992. *The Land Stewardship Letter*. Marine, MN.

Madeley, John, ed. May/June 1992. "The World's Choice: Genetic Diversity or Starvation." *International Agricultural Development* 12, no. 3.

Reintjies, Coen, Bertus Haverkot, and Ann Waters-Bayer. 1992. *Farming for the Future: An Introduction to Low-External-Input and Sustainable Agriculture*. London: Macmillan Press.

Savory, Allan. 1988. *Holistic Resource Management*. Washington, DC: Island Press.

Livestock

Board on Science and Technology for International Development. 1991. *Microlivestock: Little-Known Small Animals with a Promising Economic Future*. Washington, DC: National Academy Press.

Druning, Alan, and Holly Brough. July 1991. "Taking Stock: Animal Farming and the Environment." Worldwatch Paper 103. Washington, DC: Worldwatch Institute.

Hatcher, Gordon. 1984. *A Planning Guide for Small-Scale Livestock Projects*. Little Rock, AR: Heifer Project International.

Jacobs, Linda. 1986. *Environmentally Sound Small-Scale Livestock Projects: Guidelines For Planning*. Arlington, VA: VITA Publishing Services.

MacDonald, Ian, and John Low. 1985. *Livestock Rearing in the Tropics*. London: Macmillan Publishers, Ltd.

McDowell, R. E. 1991. *A Partnership for Humans and Animals*. Raleigh, NC: Kinnic Publishers.

McNitt, J. I. 1983. *Livestock Husbandry Techniques*. London: Collins Professional and Technical Books.

Pelant, Robert K. 1992. *Chain of Life: Passing on the Gift in Heifer Project International*. Little Rock, AR: Heifer Project International.

Winrock International. 1978. *The Role of Ruminants in Support of Man*. Morrilton, AR.

Rural Development

Batchelor, Peter. 1984. *People in Rural Development*. Exeter, U.K.: The Paternoster Press.

Bryant, Coralie, and Louise G. White. 1984. *Managing Rural Development with Small Farmer Participation*. West Hartford, CT: Kumarian Press.

Bunch, Roland. 1982. *Two Ears of Corn*. Oklahoma City: World Neighbors.

Huillet, Christian, and Pieter Van Dijk. 1990. *Partnership for Rural Development*. Paris: Organization for Economic Cooperation and Development.

Evaluation

Aaker, Jerry. 1990. "A Process Approach to Self Evaluation." New York: Lutheran World Relief. Mimeographed. Also available in Spanish.

Aaker, Jerry, and Jennifer Shumaker. 1994. *Looking Back and Looking Forward: A Participatory Approach to Evaluation*. Little Rock, AR: Heifer Project International.

American Council of Voluntary Agencies for Foreign Service. 1983. *Evaluation Sourcebook for Private and Voluntary Organizations*. New York: American Council of Voluntary Agencies for Foreign Service.

Heifer Project International. 1985. "Evaluation Manual." Little Rock, AR: Heifer Project International. Mimeographed.

Marsden, David, and Peter Oakley. 1990. *Evaluating Social Development Projects*. Oxford, UK: Oxfam.

Private Agencies Collaborating Together. 1989. *Participatory Evaluation: A User's Guide*. New York: PACT.

World Resource Institute. 1990. *Participatory Rural Appraisal Handbook*. Baltimore, MD: WRI Publications.

Zivetz, Laurie. 1990. *Project Identification, Design and Appraisal*. Canberra, Australia: Australian Council for Overseas Aid.

Manuals

Agroforestry

Folliott, Peter F., and John L. Thames. 1983. *Environmentally Sound Small-Scale Forestry Projects: Guidelines for Planning*. New York: CODEL.

Rocheleau, D., F. Weber, and A. Field-Juma. 1988. *Agroforestry in Dryland Africa*. Nairobi, Kenya: International Council for Research in Agroforestry.

Douglas, J. Sholto, and Robert A. de J. Hart. 1985. *Forest Farming*. London: Intermediate Technology Publications.

Animal Health

Hall, H. T. B. 1986. *Diseases and Parasites of Livestock in the Tropics*. Harlow, UK: Longman Scientific and Technical.

Haynes, N. Bruce. 1978. *Keeping Livestock Healthy: A Veterinary Guide.* Charlotte, VT: Garden Way Publishing.

Sainsbury, David, and Peter Sainsbury. 1988. *Livestock Health and Housing.* London: Bailliere Tindal.

Salk, Herman, and Silvia Salk. 1985. *Paraveterinary Training Manual.* Little Rock, AR: Heifer Project International.

Spaulding, C. E. 1976. *A Veterinary Guide for Animal Owners.* Emmaus, PA: Rodale Press Inc.

Werner, David. 1990. *Where There Is No Doctor: A Village Health Care Handbook.* Palo Alto, CA: The Hesperian Foundation.

Aquaculture

Bryant, Paul, Kim Jauncey, and Tim Atack. 1990. *Backyard Fish Farming.* Dorset, UK: Prism Press.

Chakroff, Marilyn. 1976. *Freshwater Fish Pond Culture and Management.* Washington, DC: VITA.

Murnyak, Dennis, and Meredith Murnyak. 1990. *Raising Fish in Ponds: A Farmers' Guide to Tilapia Culture.* Little Rock, AR: Heifer Project International.

Tillman, G. 1981. *Environmentally Sound Small-Scale Water Projects: Guidelines for Planning.* New York: CODEL.

Beef Cattle

Botswana Ministry of Agriculture. 1980. *Beef Production and Range Management in Botswana.* Gaborone, Botswana: Botswana Ministry of Agriculture.

Esminger, M. E. 1983. *The Stockman's Handbook.* Danville, IL: The Interstate Printers and Publishers.

McNitt, J. I. 1983. *Livestock Husbandry Techniques.* London: Collins.

Beekeeping

Agromisa. 1988. *Beekeeping in the Tropics.* Wageningen, Netherlands.

Attfield, Harlan. 1989. *A Beekeeping Guide.* Arlington, VA: VITA.

Clauss, B. 1982. *Bee Keeping Handbook.* Washington, DC: Agricultural Information Service.

Gentry, Curtis. 1984. *Small Scale Beekeeping.* Washington, DC: Peace Corps, Office of Training and Program Support.

Dairy Cattle

Aagaard, S. E. 1988. *Dairy Cattle Husbandry.* Little Rock, AR: Heifer Project International. (Edited and translated by Erwin Kinsey in Swahili for use in Tanzania and Uganda.)

Heifer Project International. Cow Lifetime Cards and Cattle Production Calendar Wheel. Little Rock, AR: Heifer Project International.

Heifer Project International, Tanzania. 1992. "Dairy Production for

Small-Scale Farmers: A Manual." In Swahili. Arusha, Tanzania: Heifer Project International. English version forthcoming.

Heifer Project International, Uganda. 1992. "Six Dairy Training Booklets." Little Rock, AR, and Kampala, Uganda: Heifer Project International.

Kinsey, Erwin. 1993. *Integrated Smallholder Dairy Farming Manual.* Tanzania and Arkansas: Heifer Project International.

Ministry of Livestock Development. 1983. *Zero-Grazing Series: Farm Demonstration Extension Package.* Nairobi, Kenya: Ministry of Livestock Development.

Ministry of Agriculture and Fisheries. 1985. *Modern Dairy Farming in Tropical and Subtropical Regions.* Parts 15. The Hague, Netherlands: Ministry of Agriculture and Fisheries.

Ministry of Livestock Development. 1983. *Zero Grazing Series: Housing, The Management of Napier Grass, Calf Rearing, The Fertility of the Dairy Cow, The Feeding of the Dairy Cow.* Nairobi, Kenya: Ministry of Livestock Development.

Draft Animals

Starkey, Paul. 1988. *Animal Traction Directory: Africa.* Wiesbaden, Germany: GATE.

Starkey, Paul. 1985. *Harnessing and Implements for Animal Traction: An Animal Traction Resource Book for Africa.* Wiesbaden, Germany: GATE.

Starkey, Paul, and Fadel Ndiame, eds. 1986. *Animal Power in Farming Systems. The Proceedings of the Second West Africa Animal Traction Networkshop.* Wiesbaden, Germany: GATE.

Watson, Peter R. 1981. *Animal Traction.* Washington, DC: Peace Corps.

Energy

Bassan, Elizabeth Ann. 1985. *Environmentally Sound Small-Scale Energy Projects: Guidelines for Planning.* New York: CODEL.

Intermediate Technology. 1985. *A Chinese Biogas Manual: Popularizing Technology in the Countryside.* London: Intermediate Technology Publications.

Singh J. B., R. Myles, and A. Dhussa. 1987. *Manual on Deenbandhu Biogas Plant.* New Delhi, India: Tata McGraw-Hill Publishing Co.

Forage and Feeding

Crowder, L. V. and H. R. Chheda. 1982. *Tropical Grassland Husbandry.* London: Longman Group Ltd.

Devendra, C., ed. 1990. *Shrubs and Tree Fodders for Farm Animals.* Proceedings of a workshop in Denpasar, Indonesia, 24–29 July 1989. Ottawa, Canada: International Development Research Centre.

Göhl, Bo. 1981. *Tropical Feeds: Feed Information Summaries and Nutritive*

Values. Rome: Food and Agriculture Organization of the United Nations.

Humphreys L. R. 1987. *Tropical Pastures and Fodder Crops*. Harlow, England: Longman Group Ltd.

McDowell, Lee R. 1985. *Nutrition of Grazing Ruminants in Warm Climates*. Orlando, FL: Academic Press.

O'Reilly, M. V., ed. 1975. *Better Pastures for the Tropics*. Rockhampton, Australia: Cheetham Salt Ltd.

Voisin, André. 1966. *Better Grassland Sward*. London: Crosby Lockwood & Son, Ltd.

Goats

Devendra, C., and Marca Burns. 1983. *Goat Production in the Tropics*. London: Commonwealth Agricultural Bureaux.

Devendra, C. and G. B. McLeroy. 1982. *Goat and Sheep Production in the Tropics*. London: Longman Group Ltd.

Drummond, Susan B. 1991. *Angora Goats the Northern Way*. Freeport, MI: By the Author, Stony Lonesome Farm, 1451 Sisson Rd.

Sinn, Rosalee. 1992. *Raising Goats for Milk and Meat*. Little Rock, AR: Heifer Project International.

Winrock International. 1982. *Goat Health Care Handbook: A Guide for Goat Producers with Limited Veterinary Services*. Morrilton, AR: Winrock International.

Miscellaneous Animals

Australian Freedom from Hunger Campaign. 1977. *The Water Buffalo*. Rome: Food and Agriculture Organization for the United Nations.

Blaxter, Kenneth, R. N. B., G. A. M. Sharman, J. M. M. Cunningham, J. Eadie, and W. J. Hamilton. 1988. *Farming the Red Deer*. Edinburgh, Scotland: Crown.

Grill, Peter. 1985. *Introducing the Camel: Basic Camel Keeping for the Beginner*. Nairobi, Kenya: Desertification Control Programme Activity Centre, UNEP.

Johnson, R. V. 1988. *Snail Production Techniques*. Sanger, CA: By the Author, Frescargot Farms, Inc.

Mukasa-Mugerwa, E. 1981. *The Camel (Camelus Dromedarius): A Bibliographical Review*. Addis Ababa, Ethiopia: International Livestock Centre for Africa.

National Research Council. 1983. *Crocodiles as a Resource for the Tropics*. Washington, DC: National Academy Press.

National Research Council. 1981. *The Water Buffalo: New Prospects for an Underutilized Animal*. Washington, DC: National Academy Press.

Poultry

Technical Centre for Agricultural and Rural Co-operation. 1987. *Manual of Poultry Production in the Tropics*. Oxon, UK: CAB International.

French, Kenneth. 1984. *Practical Poultry Raising*. Washington, DC: Peace Corps, Office of Training and Program Support.

Holderread, Dave. 1978. *Raising the Home Duck Flock*. Pownal, VT: Garden Way Publishing.

Kekeocha, C. G. 1984. *Pfizer Poultry Production Handbook*. Nairobi, Kenya: Pfizer Corporation in association with Macmillan Publishers Ltd.

Rabbits

Attfield, Harlan H. D. 1977. *Raising Rabbits*. Arlington, VA: VITA.

Cheeke, P., N. Patton, S. Lukefahr, and J. McNitt. 1986. *Rabbit Production*. Danville, IL: The Interstate Printers and Publishers.

FAO Technical Cooperation Programme. 1986. *Self-Teaching Manual on Backyard Rabbit Rearing*. Santiago, Chile: FAO Regional Office for Latin America and the Caribbean.

Fielding, Denis. 1991. *Rabbits*. London: Macmillan Education Ltd.

Lukefahr, S. D. 1992. *The Rabbit Project Manual*. Little Rock, AR: Heifer Project International.

Sicwaten, Juan, and Diane Stahl. 1982. *A Complete Handbook on Backyard and Commercial Rabbit Production*. Washington, DC: Peace Corps, Office of Program Development.

Sheep

Devendra, C. and G. B. McLeroy. 1987. *Goat and Sheep Production in the Tropics*. Harlow, UK: Longman Group Ltd.

Mason, I. L. 1980. *Prolific Tropical Sheep*. Rome: Food and Agriculture Organization of the United Nations.

Thedford, Thomas R. 1990. *Sheep Health Handbook: A Field Guide for Producers with Limited Veterinary Services*. Morrilton, AR: Winrock International.

Swine

Dayrit, Ricardo El. S. 1979. *Swine Raising*. Cavite, Philippines: International Institute of Rural Reconstruction.

Eusebio, J. A. 1987. *Pig Production in the Tropics*. Harlow, UK: Longman Group Ltd.

Holness, David H. 1991. *Pigs*. London, UK: Macmillan Group Ltd.

Training

Adams, M. E. 1987. *Agricultural Extension in Developing Countries*. Harlow, UK: Longman Group Ltd.

Baxter, Michael, Roger Slade, and John Howell. 1987. *Aid and Agricultural Extension*. World Bank Technical Paper, no. 7. Washington, DC: The World Bank.

Catholic Fund for Overseas Development. 1989. *Renewing The Earth: Study Guide for Groups*. London: The Catholic Fund for Overseas Development.

Hope, Anne, and Sally Timmel. 1984. *Training for Transformation: A Handbook for Community Workers.* Zimbabwe: Mambo Press.

Oakley, P., and C. Garforth. 1985. *Guide to Extension Training.* Rome: Food and Agriculture Organization of the United Nations.

Roberts, Nigel. 1989. *Agricultural Extension in Africa.* Washington, DC: The World Bank.

Srinivasan, Lyra. 1990. *Tools for Community Participation: A Manual for Training Trainers in Participatory Techniques.* New York: PROW-WESS/UNDP.

Svendsen, Dian S., and Wijetilleke Sujatha. 1988. *NAVAMAGA: Training Activities for Group Building, Health and Income Generation.* Washington, DC: OEF International.

Van Den Ban, A. W., and H. S. Hawkins. 1985. *Agricultural Extension.* Harlow, UK: Longman Group Ltd.